PEARL RIVER

James Dodds.

JAMES DODDS

PEARL RIVER

A Wivenhoe Shipwright's Story

WIVENHOE
Jardine Press

Jardine Press Ltd 2023
isbn 978-1-8382272-2-7

www.jamesdodds.co.uk
www.jardinepress.co.uk

for my father
Andrew Dodds

List of Illustrations

(Linocuts © James Dodds)

Contents

Preface

Bedtime stories from my father – *Treasure Island, Jim Davis* and *Moonfleet* – were those that first fired my soul with sea adventures. But the one that lit up my young imagination as a writer, and inspired me to set down this story half a century later, was given to me when I was just fourteen and working weekends on a Baltic trader sailing ship. John Leather's *The Northseamen: Story of the Fishermen, Yachtsmen and Shipbuilders of the Colne and Blackwater Rivers* is the maritime history of an east coast river that blends fact and fiction. I illustrated three books for another maritime writer, Hervey Benham, and I remember him wondering at the time if it was acceptable to "fill in the gaps between facts" for the sake of a good yarn. As the owner-editor of a local newspaper he had the journalist's gift for bringing a news item to life, but fiction offers more scope for elaborating a story. For him, the model for intertwining fact and imagination was *Men of Dunwich* by Rowland Parker. Dunwich was the largest medieval port on the East Anglian coast, swept away over the centuries by coastal erosion, its remains now lying under the North Sea. Time affects the same process on people's lives: the facts are gradually eroded, becoming fewer and harder to connect, except through the imagination.

J. Wentworth Day was a forerunner of John and Hervey's style of writing, famous for collecting his stories from colourful fishermen in the Jolly Sailor public house in Maldon. Interestingly, Hervey's writing advice to me was: do your research but then put it away and tell the story as if you're talking to someone in the pub who knows nothing of the subject. He advised me to return to the facts only at the end, to ensure I hadn't strayed too far from authenticity.

A worthy inheritor of Benham's maritime writing is David Patient. His book, *One of Howard's*, tells the story of John Howard, a pioneering Maldon shipbuilder in the late nineteenth century, and it combines his practical shipbuilding knowledge with the ability to tell a good story. I served my apprenticeship with David learning the skill that had been directly passed down from John Howard's men.

I cannot claim to have these writers' storytelling abilities. My own yarn about shipbuilding in the eighteenth century is partly inspired by surviving that early shipwright apprenticeship in the 1970s, in a yard where most of the work was still done with hand tools and wood, using skills little changed since the building of the Ark. I was still an apprentice when I illustrated some of Hervey's books, and later when I was an art student at Chelsea and the Royal College I was co-author of a book, *Building the Wooden Fighting Ship* set in the Woolwich Royal Dockyard in 1760. And forty years later I collaborated on *River Colne Shipbuilders: A Portrait of Shipbuilding 1786-1988.* Both projects involved combining my practical knowledge of wooden shipbuilding with academic research, firstly at Greenwich and Kew, then lately at Nottage Marine Institute and West Mersea Museum (Benham and Leather archive) and online public records.

This story is about a fictional eighteenth-century shipwright living in my home town of Wivenhoe. The way I created it was a bit like the making of a rope. I teased out the facts from the fibres of eighteenth-century wills and parish records, combining them with my own memories and imagination. I spun these three strands together by walking backwards through history, twisting many yarns into a story. With this rope the sails of the story were raised, binding together a family, a ship's crew and a community.

So here is that story. Most events, names and dates are real, in tribute to the people whose lives I've enjoyed uncovering; their stories have been spun afresh with the help of my fictional William King and his son.

James Dodds

Part One

William King

1
William King

My father, a shipwright, told me I was born in 1714, the year of the new King George. Little King William he called me, since I was named after him, William King. He built ships with Taverner, Austin and the two Johns, just a stone's throw from where we lived. I have no memory of my mother Hannah, she died a year after I was born. I was brought up by my father and Aunt Mary in my grandmother's house.

My grandmother Mary King was a fine old lady. My earliest memory is of sitting at her knee toasting pikelets on a long fork in front of the kitchen fire. She'd tell me stories of our family history, of the shipbuilders and sea captains that built the town. Of our ancestor the brave Captain John King who fought the Dutch, and his son Francis who married Hannah, daughter of Robert Page shipbuilder of the famous *Nonsuch*. Francis was master of a pink called *Success* who traded with Rotterdam. He got in trouble for not paying duty on some books he'd shipped for a Scotsman, believing what he'd been told, that they were just ten parcels of plain paper.

Grandmother had the good fortune to inherit the copyhold of the Woolpack inn and other properties from her wealthy uncle, Jonathan Feedam. A good and charitable man, he and his brother James were mariners who made their money with their ships, Rotterdam Merchant and Jonathan, making regular trips to the low countries. Their sister Ann married another of Captain John King's sons, William.

PEARL RIVER

Grandmother said we were from the trade side of the family. Her husband Robert was a ship's blockmaker. He leased the woodyard, which was part of the shipyard. He also had a tenement in a house called Hartshorne in Hogg's Lane, or West Street as it is called now, along with the copyhold of the Woolpack, now called Ship at Launch; we just call it the Ship.

It was the two Marys, my grandmother and my aunt, also called Mary King, who looked after me in in my early days, but I could not wait for Father to come home from work and tell me his stories from the yard. He would talk reverently about the men he had worked and learnt his trade with: shipwrights Robert Archer and William Barnsley, and his own father, another William King. "'Tis an unbroken line," he would say, "that stretches back to Noah or even further: to the very first man to cut and sit astride a tree and float on water."

Father had a cutlass he said belonged to Captain John King, and I would beg to see its fine Spanish steel and hear the story of how this brave ancestor came by it. John had gone to sea when he was twelve, a cabin boy on a merchant ship that was attacked by a heavily-armed Spanish privateer on its way to the West Indies. The crew were put in chains and given the choice of joining the Spanish or being put ashore. But John managed to slip his manacles, the chains being made for a grown man's ankles, and escaped while most of the Spanish crew were ashore trying to catch wild pigs. He stole this very cutlass, cut the anchor cable with its strong blade, and in the confusion was able to steal the key to the manacles and free his shipmates, overpowering the few Spaniards left aboard. They commandeered the Spanish ship and John's captain showed his gratitude by ensuring John was well looked after for the remainder of the voyage. He personally tutored the lad in using the quadrant and all the navigational skills he'd need to become a captain himself. Having sailed out with a leaky old merchant ship, they came home with a fine Spanish prize and the crew got a share of the prize money, after the merchant shipowner had been compensated. John and his eight-gun ketch *Nonsuch*, built by Robert Page, were hired by the Navy "to work among the sands": his first command. He went on to be captain of the 26-gun ship *Mermaid* at the Battle of Gabbard for Cromwell and the 46-gun ship *Diamond* at the Battle of Lowestoft for the King.

My father had many stories. The one that afeared me the most was about the black dog Shuck that haunts the marshes. A foolish fellow had been out on the marsh after dark, in search of some casks of brandy. He'd overheard two fishermen talking about where they were hidden, at the mouth of a small creek. But he lost his way and came face to face with the burning red eyes of an enormous black dog. He knew the legend: if you saw it you'd meet an untimely death. He was so shaken that he ran straight into the mud, up to his knees and stuck fast. Hearing his cries for help, the villagers went searching, but could not find him until the next day, after the tide had drowned him. Father would tell me: "Never venture out on the marsh after dark!" And if I was ever to get stuck in the mud: "Never fall for'ard. Fall onto your back, spread yer arms and ease one leg out at a time, then slide yourself on your back to harder ground."

Then there was the couple that lived alone on a strange island with a freshwater spring and a pond in the middle of the salty marsh, up behind Mersea. The woman got so lonely when her husband was at sea she would invite strangers and fishermen to call on her with news of her family. When her jealous husband got to hear, he took her out in his boat, chained her to the thwart and smashed a hole in the bottom with an axe. Drownded the both of them.

Grandmother did not like to hear Father telling me these stories. "The world is full enough with horrors!" she said. "Do not be filling the poor boy's head with such tales." He would give me a wink, which made it all the more special between us.

My father's sister Mary was always busy around the house and had little time for stories, but she would sometimes read me a parable from the Bible. I liked the story of the flood with Noah and his sons building the Ark. And the tower of Babel and why the world has different races and languages. Occasionally she told me about my mother and how lovely she was, and then it would be back to her work around the house.

When I was eight, Aunt Mary married a surgeon called Horace Flack. I liked Horace, he was very kind to me. He'd tell me stories written by Homer: of Ulysses and the golden fleece, the one-eyed giant and the mermaid sirens. The line "Athwart the fishy flood," has always stuck with me. Horace was also patient enough to teach me to read and write, a skill I am eternally grateful for. He said: "Books will open worlds to you."

Horace was famous for establishing a saltwater spa behind the Woolpack inn and had that grand seashell porch put over the doorway there. Water was let into a tank in the stables, the mud settled out and then the water was let into a big bath. People paid one guinea for the season or a shilling for a bathe. Good money to dip into salty river water! But the spa was a great success: even Mary Rebow from the big house in the park enjoyed a visit or two. It was something I could get for free in the river if I cared for it, and Horace was convinced cold water bathing was good for you. It didn't do him much good. In the end his limbs got so weak he could no longer be going after it himself.

The Woolpack was next door to the customs warehouse, that later got known as the Vice Admiral's Quay. The warehouse as far as I could see was full of old anchors and parts of wrecked ships salvaged from the sea, but I heard it also held seized smuggled treasures, from silk and lace to tobacco and tea, brandy and the Dutch genever*.

Horace gained even more respect in the town after fixing up some soldiers that got hurt on board Captain Martin's cutter *Wivenhoe* after a fight with smugglers. The smugglers did not even bother to hide when they landed the contraband. They just relied on outnumbering the revenue men.

Horace's surgery and apothecary occupied the front room of our house, and it was a grand sight to see all the jars lining the walls. The door was always kept locked. I could not understand why so many poisons were needed to cure people. I remember Horace being called up to the Hall to bleed a baronet called Sir Caesar Child, who had lost his money with that South Sea company. He was staying with his son-in-law, Nicholas Corsellis** at the Hall. Horace said it was as if these families saw themselves like the Romans, slave-owning empire builders***.

---

* *Dutch courage*, or *genever* (Dutch for *juniper*), otherwise known as *gin*.
** Nicholas named his son Nicholas *Caesar* Corsellis.
*** The Caesars, another family connected to the Corsellises also lost a lot of money. Sir Child Caesar (not Sir Caesar Child!) was imprisoned for debt in the 1730s.

He said slavery was an abomination and Child deserved to lose his money. "Money corrupts money," he said. Many politicians were found to have taken bribes to change the laws giving the South Sea company a monopoly on trading slaves to South America. I did not at the time fully understand what he meant, but I now do. Even after losing all that money the company still carries on. "It's too powerful to fail," Horace said. It seems to me the stock market is just a fancy gaming house. The state lottery is no better, creating more misery than good. Sometimes I wonder how things would be if the old Commonwealth that Grandmother talked about had prevailed, with their laws banning gambling.

When I was old enough I used to like helping Father out at the shipwright's yard. It always held an air of excitement for me. Smells of mud, tar, damp timber and tools, all this contained a magic, a different sort of alchemy to Flack's surgery. The shipyard was where, with the collaboration of men and materials, a ship was made that could sail the wide world round, discovering new lands and treasures to plunder. I knew the other shipwrights did not like me hanging about, getting in the way as they saw it. They would tell me: "Go home! This is a place for men, not for a child," which made it all the more fascinating to me. Home was filling up with Horace and Mary's children so I was spending more and more of my time outside.

The yard and the river were my playground and education. We fished for eels and grubbed for flounders at low tide. My friend Tom and I were inseparable. Tom had from a very early age sailed with his father on his oyster boat. His father let us use a leaky old skiff. We rigged it using an old oar for a mast through the fore thwart, and another for a yard, and we made a tiny standing lug sail out of what was left of a rotten old sail we found in the yard, set well for'ard with a loose foot. The main sheet just fed through the ring on the stern post and we steered with an oar. Together we explored the river Colne upstream to the lock gates at Colchester Hythe, or took the Roman river to Fingringhoe mill. We explored the maze of little channels, following the flow of the tide, in and out of the reed beds as we pretended to be brave Captain King and his trusted lieutenant, hunting pirates on the mighty river Orinoco or hiding from Roman soldiers on the Tigris. Sometimes we would sail down to Alresford creek and follow it to Bricklesea* Church Quay where the town had been before the Black Death – you can still see the dents in the field up to the church where the houses had been – and further up to Thorrington Mill. Going downriver meant sailing against the tide at some point or another but we could do it if the wind was right. Any further downriver we'd have struggled to get back. Many a time we ended up sculling or getting out and pushing the boat. Once we pulled the boat ashore and walked home covered in mud. We got such a scolding. We walked back on the next tide to get the boat. All great fun! Much later I built a fishing sloop for Tom, who had taken over his father's oyster laying and pit.

When my father got ill, Horace did all he could. First it was thought he had the ague, but it was smallpox, and none of Horace's remedies could save him. They would not allow me to see him. We were all very afeared of catching the pox,

* *Bricklesea* is now called Brightlingsea.

Mary particularly for her children. Horace said most people recovered, but he had no cure. "All we can do is keep him comfortable and warm and try encourage him to eat, drink and clean his sores," he told us. Only the maid, who'd survived smallpox when a child, was allowed to attend him. We all prayed for his recovery, but I knew it was no good when Vicar Goodwyn called to see him. It was the 3rd July, 1726, when we buried Father. Horace sort of took his place as head of the family, but Grandmother was still very much in control.

My father left the shipwright's yard to his sister Mary. I was only twelve, she was not long married, and, like her mother, good with money. She could collect and pay the yearly rents to the Manor. I guess he thought it would be more secure in her hands than mine. I thought at the time that he had left it in her safe hands for me to take over when I was old enough. But that was not to be. It always seems it is the women in our family that hold the property, the silver and the purse strings.

Two years after the smallpox had taken Father, my official apprenticeship started. I had just turned fourteen and was indentured to John Iffe "to be instructed in the art of shipbuilding for seven years." Grandmother paid him and the stamp duty.

2
Apprenticeship

I did not get off to a good start with old Iffe. The first thing he said to me was: "I am stuck with you now, I shall not tell you anything, you will just have to watch me." He was only twelve years older than me, but seemed more. I thought I knew more about the shipwright's trade than I did! Just from being my father's son and watching him about the yard. I now know that watching is just the start of it; doing is where the real learning is done. John did not like it that my father's sister, Aunt Mary, had inherited the copyhold of the yard, and the fee for my indenture had probably been much reduced because of it. Iffe had to pay her rent while she, of course, still had to pay yearly rent to the Corsellis family in the big house. They were a Dutch family who'd settled in Wivenhoe about a hundred year ago and owned most of the land and properties in Wivenhoe.

Anyway, if I did something wrong with John Iffe it was always: "What would your father have say to that?" I felt like the whipping boy, getting the blame for everything, even for putting that rotten fish in his new sea boot, which must have been Robert Clear, the other apprentice. Robert was the son of an oysterman, younger than me, a handsome boy with sharp features and dark eyes. He would come to work in his old brown wig and long white waistcoat and frock coat, thinking himself quite a dandy, and he was full of stories. We got on very well. We would sometimes have a small beer in the Maidenhead along the quay away from the shipyard and he would make up some fanciful story about us running away to sea and returning home wealthy men, with a parrot or a monkey on our shoulders and beautiful foreign wives on our arms. He was always telling me about some young maid or other who had caught his eye. Then he just disappeared. He advertised in the *Ipswich Journal* for him to return; he still had four years of his apprenticeship to serve.

I missed him.

My first job at the yard was just sweeping the shavings and adze dubbings from around whatever the two Johns were working on. If it was oak I had to sack it up for the smoke house. They got a penny or two once a year for the results of my toil.

I always had one eye on the river. All sorts of ships: snows, pinks and hoys* carrying everything from coal, timber, wool and spirits. The pinks can sometimes get all the way up the river, but the real excitement happens when the bigger ships come up on the tide and put their bow into the mud of marsh opposite. The tide then swings the ship around. A boat takes a line across and they warp them over to the quay to unload into the warehouses or lighters for Colchester. The lighters are used to take goods up river where it is too shallow for the ships. The lightermen row them up, arriving on the top of the tide to get though the lock gates where they can remain afloat, returning on a following ebb after unloading and loading with whatever is to be shipped out. When the ships are reloaded they catch the next ebb tide out into the wide world. The packet boats

* A *snow* is the largest of the two-masted ship-rigged merchant ships. A *hoy* is a single-masted cargo ship with a sloop rig, like a cutter. A *pink* is a shallow-drafted flat-bottomed cargo ship with a high pinched stern. They are all merchant ships.

leave twice a week to London with their cargo of bays, says and perpetuanas* from the Dutch weavers in Colchester and return with their raw material of wool. But the real treat is seeing the beautiful revenue cutters setting off on patrol or returning home with stories of fights and seized contraband, and sometime a prize ship.

One day an announcement in the *Ipswich Journal* read: "To be sold to the highest bidder, by the inch of candle from her Majesty's warehouse in Wivenhoe: 164 small casks containing upwards of 640 gallons of foreign brandy." There was much excitement in the town come auction day. I thought: "Perhaps if I ever build my own little sloop one day I can catch myself a cask or two."

It is not only contraband that is stored in these warehouses but also goods salvaged from wrecks, for which the Thames estuary is well known. A ship can go aground out of sight of land and be pounded to pieces on one of the many sandbanks hereabouts. Salvage is considered fair gains and an abandoned wreck is often plundered by the fishermen. By law they have to hand the goods over to the receiver of wrecks. The goods are auctioned and a third goes to the original owner, a third to the Lord Warden of the Cinque Ports and a third to the fisherman who risked his boat and his life to bring it ashore. You can imagine a few items never make it to the auction.

Iffe would keep calling me back: "Stop your dreaming boy, keep your eyes and thoughts on the job!"

---

* *Say* is a light serge: thin woollen stuff of twilled structure. *Bay* is similar to baize but lighter in weight and with a shorter nap. *Perpetuana* is woollen cloth.

It was not long afore Iffe let me have a go with an adze. I had my father's tools and thought I knew how to use them. Father had sharpened them and carefully wrapped them in oily cloths in his chest. It must have been the last thing he did afore he died, or perhaps his brother Samuel done it? Sam was a house carpenter and understood the value of tools. He died the year after my father in 1727, same year that George II and Queen Caroline come to the throne. However, Iffe said my father's adze could do with a bit of a touch-up. He said it should be sharp enough to shave the hair off your arm. He showed me how to hone the blade with a whetstone in a circular motion and how to slip out the handle to grind the blade on the grinding wheel. That was probably the first real lesson Iffe gave me. "Keep your tool sharp! 'Tis safer to have sharp tools." Iffe told me of a shipwright that could split a penny in half, edge to edge, with one blow. "Though it wouldn't do the blade much good, or the man," he said. The same man could carry a cutter's main sail on his own back!

The two Johns, Iffe and Davis, had got the job of making a new topgallant mast for a cutter. They kept the trees for masts in the river. I would sometimes help the scavelmen keep the beach clean by working on the falling tide with long wooden hoes called hummers, so called for the vibration you feel up the handle. That way the stony beach is kept clear of mud, the same way they keep the ford and graving bridge clean. Old Flack had to make sure he only filled his bath on the floodtide!

But we needed help to pull the tree up the beach. We used a "pair-a-wheels"* that we rolled over the tree and secured to

* A *pair-a-wheels* is a pair of cart wheels on an axel with a long shaft.

the leading end of the log with chains. Then we pulled the long shaft down, to raise that end of the tree off the ground, hitched up a horse and pulled it up onto level dry ground next to the graving dock.

Iffe explained we had to make the round tree square before we could start making it round again. In the Royal Dockyards up the Thames there they have their own mast ponds. Keeping the trees in the saltwater helps preserve them and stops them drying out too quick and cracking up. Iffe told me they have mast-makers that will do nothing other than make masts. In the King's yards nobody builds a ship from start to finish, it is all broken down into different trades. But here in Wivenhoe we get to do a bit of everything. He told me that when he was a boy he visited Harrison's new ropeworks up at the Cross and got into trouble for not leaving his pipe and baccy at the gate. There is always a chance of fire, though most of the ropemaking was done outside, only the store and crank wheel are under cover. He said if you got caught with a pipe or tinderbox in one of the King's shipyards where they do all the work inside a building a quarter of a mile long, they will hang you for treason! So important a job for the Royal Navy is ropemaking, and so afeard of fire are they. He could see me taking an interest. I think that's when we started to get on, getting the measure of each other. Though 'tis hard to imagine old Iffe was ever a boy.

Anyway, we blocked the tree up to work on and held it in place with wedges secured with timber dogs. We found the centre of the tree rings at each end and plumbed a line, this gave us a line to work from. The mast had to taper, a bigger diameter at the heel. We drew a nine-and-a-half-inch square at the heel and a four-and-three-quarters-inch square at the head, scribed the square to the edge of the tree and stretched a string line between. When it was tight we ran a lump of chalk along the string, stretched it even tighter, and then Iffe pinged the string against the side of the tree. It was 36 foot, which looked long to me, but Iffe said her main mast is more than twice this long at 80 foot, with a 22-inch diameter at the heel. Iffe told me a man-o'-war's masts are massive and made in pieces, held together with iron bands.

We marked the sides first. This gave us two lines to cut to with a handsaw. Then we chopped the blocks off with an axe, and adzed from one end to the other. At first I found it hard to get the knack of getting the adze blade always to hit in the exact same place but Iffe showed me the trick. He could swing the adze above his head and get it to fall on the same spot each time. It would be a long time afore I could do that.

I sometimes got to walk down the river, to where the creek cuts off to the ford and the tide mill, collecting lumps of chalk to use in the yard. If I was lucky, I got to use the boat but I had to judge the tide right. I could scull down with the tide, find a few lumps and catch the flood tide back again. Sometimes we got chalk from one of the south coast boats bringing a load up for the oyster layings, good for the making of shells, I guess. It also came in handy after days of caulking a deck, you would get caulker colic being bent double day after day, and chewing a bit of chalk would help.

It took us all over a week of chopping and planing to get the tree square. Iffe showed me, by putting your eye close and looking along the pole you could see any lumps and bumps that needed planing down. I soon got my eye in and was telling him where to plane a bit more off. It was only when Iffe was properly satisfied all was straight that we divided up the four sides into eight with the chalk line and removed the wood triangles. Then it was divided into sixteen, and so on. More and more of the work was done with a draw knife and long planes. In truth the two Johns were doing most of the labour, I was still just the boy, but I did my best to keep up. My muscles had never ached so much. We had an enormous pile of kindling and firewood for sale or to take home, all tied up in bundles or bagged up by me, though it did need drying out. I am told some people call us chippies because a chip is the size of leftover wood that shipwrights in the big shipyards are allowed to take home for their fires.

Iffe told me the story of how Captain Matthew Martin made his fortune commanding the 480-ton East India Company merchant ship *Marlborough*, with a crew of 96, and 32 guns. In 1712 the *Marlborough* was attacked by three French ships and fought them off for three days. Our good captain deceived his pursuers at night by setting a cask adrift with the ship's stern lantern on it, giving the *Marlborough* time to escape. He was richly rewarded, married one of the East India company director's daughters, became a director himself, then an MP, and finally mayor of Colchester. There were a least two other East India company Captains with a Wivenhoe connection, Captain James Kettle who married Elizebeth Corsellis, and later Thomas Best, who became a captain of the East Indiaman *Prince Henry*. Thomas is the same age as me. His first wife was Captain Francis King's daughter Hannah.

As my apprenticeship progressed I was able to help with the yard's main work, which was mostly building tuck-sterned sloops* and smacks between 27 and 58 foot long for the local fishermen. The most common size was around 32 foot.

Between the building work our time was taken cleaning ship's bottoms beached on the hard ground between Anchor Hill and Quay Street, or when the tides were right, in the graving dock. We breamed the bottoms of all sorts of vessels. The revenue cutters were the most regular, for a clean bottom is very important to them. We would pull the ship into the graving dock, sitting her on beams set into the floor of the dock. We needed a good tide, the best being around midday following a full moon. We'd haul her over on one side as the tide went out. Scrape and burn off the weed and barnacles. We burned reed faggots held to the ship's side with a breaming fork, taking it in turns to scrape or hold the fork. Some think charred wood gives protection from the worm. Quite often a bit of caulking was required here and there, and then we would give the bottom a fresh coat, a hot mixture of rosin, tallow, linseed oil and a bit of brimstone. The modern revenue cutters now have copper bottoms, sheets of copper nailed onto a mix of cow hair and tar. These do not need graving so often and it is lot easier to clean them off, but it makes replacing a plank or caulking a bit more complicated, you have to get the copper sheets off first. It was all very dirty work. The next tide we would lean her the other way and do the same all over again.

---

* *Sloop*: A vessel with a single mast and a fixed bowsprit. A *Smack* has a bowsprit that can be run in. A *Tuck Stern* has a small transom with a tuck or counter extending aft with the stern post and rudder passing though it.

Iffe said my father had told him how, in the old days, to build a large ship in the graving dock they would put a pair of gates across the mouth of the dock and set long poles into the ground around the ship to build a scaffold as the ship grew in size. The pair of gates would be floated into the mouth of the dock. When the tide dropped the heels of the gates were tied off to the retaining posts and after the next high tide the gates let the water out through paddle holes, then closed them so that on the next tide the force of the water would lock them tight, and then the gates would be secured with shores and any leaks would be caulked up with clay.

3
Home

I remember how lovely it was coming home in those days, tired, face aglow, hair all stiff from sweat and salt air and from a day's work by the river. The kitchen would be warm with the sweet fug of cooking. They would all be eager to hear about my day and for me repeat the stories and gossip from the shipyard.

Aunt Mary would be at the stove cooking our supper, Grandmother in her Windsor chair by the fire, knitting or reading aloud from the broadsheet with the help of her quizzing glass, which she kept on a fine chain about her neck at all times. Sometimes I would distract the young ones with one of my father's stories or one or two of Horace's *Aesop's Fables*, while their mother got on with the tea.

She would sometimes cook me my favourite meal: salt cod in a butter and egg sauce, with a few new potatoes and carrots. What a feast! She gave me the recipe once, saying one day you can ask your wife to cook this for you.

"To make Egg Sauce for a Salt Cod:

Boil four eggs hard. First half-chop the whites, then put in the yolks, and chop them both together, but not very small. Put them into half a pound of good melted butter and let it boil up. Then pour it on the fish."

Then she would try and get out of me if a girl had caught my eye: "Are you not sweet on anyone?"

"They have to cook like this first," said I.

Looking back, when Aunt Mary died I felt I had lost my mother again. No one to fuss over me when I cut my hand or cook me my favourite meal. I was a married man by then but I still felt it as a great loss. Horace grieved but got himself remarried soon after to James Feedam's daughter Leah. He was not the first in the family to marry a Feedam. Feedams, Martins, Rebows, Pages, Parkers, Locks and Kings are all related to each other in one way or another!

Grandmother died at the good old age of 90 in 1746, a year that is firmly set in my head. She was a kind old soul, left me some of her silver. She bequeathed the house to Horace and everything else to Aunt Mary's children, Aunt Mary being already deceased. The eldest grandchild Anna, my cousin, inherited the copyhold of a house over the river in Fingringhoe and much besides. Anna also inherited everything else eight years later from her father Horace.

But I'm getting ahead of myself.

4
Tides of Fortune

At one and twenty my apprenticeship finished. I could now earn a shipwright's pay, legally take strong drink, and get married. I'd had my eye on Mary Hall, the oldest daughter of master mariner William Hall, for some time, having first seen her at church. Never in my life have I been so eager to hear the church bell calling us to Sunday worship. I enjoyed the singing in church, not so much the sermons. There is no better expression of faith than the individual voices of our community singing with one voice, save perhaps the building of the great cathedrals, celebrating the Maker of all things. I have always liked the idea that the "nave", the main body of the church, comes from the Latin for ship. But my thoughts at this time were on Mary.

We finally met properly as we left the church after worship one Sunday. I wished Mary and her parents good morrow and then became tongue tied, my face turning red. Mary could see my awkwardness and asked would I like to have a quick turn round the village afore she joined the pony and trap home, looking at her father and mother as she said this. They nodded their consent and her father said: "Do not be long! And take your brother with you. We will be waiting for you in the trap." They got busy talking to others from the congregation and let us slip away.

I asked Mary if her father approved of me. She replied: "He liked your father and knows you are a good worker. He has noticed you looking at me in the church over the last few months with your faraway eyes". She said she had been hoping to have the opportunity to talk to me before now but I always rushed away. I felt my face flush again and did not know what to say. She laughed, a kind laugh, and said it was all right, we are talking now. Well in truth she was talking, and I was glad of it. She had a very easy, down-to-earth manner. She was good in the art of conversing and, as I was to discover, well-read. She seemed to know her own mind, and had opinions on most things.

We talked as we walked down Love Lane, across the brook and a little way up the hill into the woods. The bluebells were as vibrant as her eyes. I said: "You can get a good view down the river from here," a view I had seen many times afore, but today it looked clear and sharp through the trees across the big sweeping arch of the river down towards Mersea island. It was such a fine clear day. I said for the lack of other words: "If it wasn't Sunday you would see a fleet of masts from the oyster boats working the layings, up the Pyefleet and the Geedons." We could just make out a few masts of pinks, hoys and maybe a snow over Bricklesea marsh, anchored up behind the Mersea stone, waiting for the morrow's tide with their cargoes for Colchester. Somewhere not too far off would be a customs cutter.

But this was not where my interest lay that fine day. She was so pretty with the sun on her pale skin, her rosy cheeks and her red hair framed under her starched white lace bonnet. Her hair tumbled about her bare shoulders. It was only the tugging of her brother's hand in hers that brought us back.

We had several Sunday walks like that, finding we had more and more to talk about, until one Sunday her father asked me to join them for Sunday dinner after the church service. It was wonderful squeezing into the back of the trap with her, as the pony pulled us up the High Street to their home on Wivenhoe Heath. The finest beef and roast potatoes and vegetables had been prepared for us. I really did feel like a king that day – that is, until her father asked me to join him in his study for a glass of brandy.

We were soon wed at St. Mary the Virgin, another very fine day. Mary wore a green silk dress she had made with her mother, with a lace veil held by a tight ring of forget-me-nots, the blue to match her eyes, and a single pearl earring.

I wore my old father's best brown frock coat and embroidered waistcoat – Grandmother said he had it made for his wedding – and a silk shirt with a lace collar and cuffs. The breeches Aunt Mary made for me. My wardrobe was finished off with some new silk stockings and buckled shoes. After the ceremony we all went to the Blue Anchor for food and drink. An old sailor played his fiddle. The wedding cake was made with savoury meat and sweetened mincemeat. They said a small glass ring was buried inside and the maid who found it was the next to be married. I am not sure if the ring was ever found, I did not pay much attention to that!

My new father-in-law, William Hall, had done well at sea. He and his wife Mary owned a small estate called Sayer's Grove on Wivenhoe Heath and also a couple of brick-built tenements on the quay, one of which we took for our home.

We were so happy, and "heel-over-head" as Grandmother used to say. Mary was strong and practical, a good cook and housekeeper. So happy until we lost our first child. We had named her Hannah after my mother. It was a great sadness. Horace told us it was not uncommon to lose your first...

There followed some difficult years. Aunt Mary died in 1739, and we lost another baby, Mary we'd called her. Only King William survived, he was born in 1743. And my poor beloved Mary died in the birthing of him. We had only been wed eight years! There was nothing Horace nor his assistant Thomas could do to save her.

Horace found young Will a wet nurse. Mary's mother and Horace's new wife Leah were very kind to us helping out with the baby, Leah just having had a child herself. This was a sad time, history had repeated itself, mothers dying. It was the first time death really hit me hard. She was a good wife and would have been a great mother given the chance. I missed her dearly and understood for the first time what my father must have felt when he lost my mother. I felt the loss of my parents all the more keenly too. I had only known the female company of my grandmother and aunt and cousins, but Mary was so much more, the finest companion and my best friend. The only thought that kept me going was that I now had to do my best for young Will.

Grandmother Mary King died in 1746, she hardly got to know young Will, and the same year my Mary's father William Hall died. Mary's younger brother John was apprenticed to an engraver in London. He made a great success of it, became historical engraver to the third King George. Sadly my beloved Mary and her father were not here to see it.

I was now in my thirties. John Davis had died when young Will was still a baby. He had only just taken on a new apprentice a year before, so James Harvey came and worked for me for the rest of his apprenticeship, finishing in 1751. John Iffe lost his wife Sarah, he was still working in the yard.

Horace died at the age of fifty-eight. I remember looking at all his books on the shelves, some in Latin, and thinking: "All that knowledge was in his head and now 'tis gone." We would sometimes sit and talk late into the night about what he thought of this book or that, or which book I should next read. It was Horace who had taught me to read and look deeper into things. He would always be recommending something or other for me to read. Often our discussions would end with me thinking that a good memorable story was the best way to pass on ideas, wisdom and knowledge. Far better than some dry book of facts.

Seeing his books after he was gone was the saddest thing. They were all dry to me now he was not there to share his favourites with me. He'd been like my second father. Leah had died a couple of years earlier, aged fifty-one, and he had not been the same since. Just before Horace died he tried to let the old shipyard, graving bridge and our old house for his and Mary's son John, who had inherited it from his mother when he was only five and had never taken an interest in it! After Horace's death William Wyatt got interested and bought it from John. Wyatt was only two years older than me. By rights I had a feeling the yard should have come to me, what did cousin John know of shipbuilding? I wished Horace had not felt so duty-bound to his real son and Aunt Mary's legacy. But that was an unfair thought, and I never voiced it. If only father had lived longer, or I had been older with an income, the shipwright's yard would have been mine. It has always been just out of my reach!

At this time I was feeling very alone. I was working by myself, struggling to find enough work and missing my dear wife. From then on it was just me and young Will.

A year after Horace died his daughter, my cousin Anna, married Horace's assistant Thomas Tunmer. Anna herself did not make old bones, dying three years after the wedding. With Anna and Horace gone, Thomas sold our old family house and moved himself and the apothecary into late Mrs. Boice's house next to Captain Harvey's. Dr. Tunmer is credited with some of the first treatment to prevent smallpox. If only they had know about this in my father's day, how different things would have been! I am told Dr. Tunmer's treatment saved many lives.

The years pass.

Sometimes I like to have an ale and a pipe in the Ship after work. The Ship is a bit upriver from the town, out of the way and a good place to be still after a day's hard work. Watch the sun setting and lighting up houses along the quay from the bay window. Watch the shadows grow. On a still autumn eve with the tide high, you can see the whole quayside dwellings and warehouses reflected upside-down in the water. Seeing the town in this different way somehow makes the worries of the place less important. Turns things on their head.

One such evening I met my old friend, the boy who had run away, Robert Clear. Well, he was a man now. I hardly recognised him, he had lost the old brown wig, but finally I could see the boy behind his pox-marked face. He had just got out of Chelmsford gaol after an uncle had paid his share of the duty on the brandy seized from the ship he was on. He told me he had got fed up of work with Iffe.

"I had to get Clear," he said with a smile.

And although as an apprentice he was not meant to drink he had overheard talk in this very building of a sure way to make lots of money. "Had not all the property owners in this town made their money at sea?" he said.

He left with the crew of the *Good Fortune* on the very next tide. "It does not look like it was so good for him," I thought. But he spun a fine yarn of his adventures, outrunning the custom cutters, fighting with the revenue men and salvaging wrecks. He had even crossed the Atlantic a few times in legal employment. Learnt the ways of the sea. And its tarry language, "Tar-appalling!" Another smile.

He told me: "You never forget your first night's sail crossing the Channel, seeing the sea sparkling with phosphorescence in your wake and the cold sky slowly lightening as dawn approaches, with the unbroken horizon completely encircling you."

His first voyage was over to Ostend to load brandy and genever. On the return journey they stood off the coast until nightfall and then, with the wind offshore, anchored near the beach and ferried the barrels up the beach, "rowing in backwards with a line aboard to ease us through the surf without being tipped over by the breaking waves. Also we used the line to pull ourselves off and find the ship again in the dark." Once on the beach the barrels were rolled up over the shingle bank and down to a small brook. The Gunfleet I think they call it, that flows through the shingle. There were men waiting in flat-bottomed punts to carry the casks inland.

"We had nearly finished our unloading when we were discovered by a troop of revenue men! There on the beach a fight broke out. All the men were armed. A few shots were fired but we easily outnumbered them and gave them such a hiding. They retreated and we got back aboard and out to sea as quickly as we could."

"You were lucky!" said I, though "Mad fool!" is what I thought.

None the less, he did say he thought I'd done the right thing to stick it out with old Iffe, learning a good shore-based skill, a trade. All he was good for was working at sea. I told him of my plan to build a little tuck-sterned sloop, if he wanted to come in with me? He said he had no money and what money he earned would go to his uncle. Already he had signed up on a timber ship bound for Archangel, up on the White Sea. I had not heard of timber being imported from so far north before. Plenty from the Baltic ports and some fine timber from the Americas but not from Archangel. I could see Robert Clear would not be content with a life ashore.

It makes sense to build ships in America rather than shipping the timber here to build them. It is only really worth importing the big tall firs for mast-making across the Atlantic. If it was not for young Will I might have gone and built a ship in New England. They say they have an oak there that is cannonball-proof!

Robert's parting story was that he was off to find a warm Pacific island full of beautiful, comely, brown-bodied women wearing nothing but grass skirts, with breasts as round and firm as breadfruit for all to see... or perhaps a rich Spanish widow, or even both! His grin could not have been wider. I said with his luck a school of mermaids would more likely lure him onto the rocks.

"I don't give a fish's tit!" he laughed. "Any mermaid or widow will do."

5
Building our Oyster Sloop

My chance meeting with Clear haunted me. I had to admit to myself that I felt an envy for his free and adventurous life. 'Twas time for me to take a risk with something new. So I set my heart to building myself a boat, making my fortune with the oysters and escaping the town like my old friend. Besides, it would give me a chance to pass on some of my skills, and a boat, to young Will.

I had some money from selling the few bits of silver that Grandmother King had bequeathed to me, and young Will had inherited some money from Mary's father, so I thought the time was right to build ourselves that little sloop and have a go at dredging oyster – maybe a barrel or two of brandy would come my way. It was time to try my luck at sea!

The first step was to buy the materials, but we had been at war with the French in the American Colonies for a couple of year now and wood was expensive. Fortunately I had bought a log of Danzig elm before the price increased. There was enough timber in that log for the keel, stern post and most of the planking. I'd been keeping it in the river. We hauled it out and removed the bark with axe, bark spade and draw knife. We used the shipyard pit, manhandled the log onto wooden runners over it and secured everything with timber dogs. Marked our first cut with the string line and chalk for me to follow. I stood on top of the log and steered with the tiller of the saw, and young Will was under the log in the pit, the underdog. He did not much care for this job, kept complaining about the dust in his eyes! But he was good and stuck at it. The wood still being green, the dust was not so bad. I would tap a wedge into the cut from time to time to stop it closing up and nipping the saw. We first cut a six-inch-wide board from the centre of the tree, rolled each side of the tree over so the flat side was facing down and ripped the two halves into inch-and-a-half boards for planking, with a couple of two-inch boards for the hog and wales. Fresh cut elm smells a bit like cow shit, but it is a wonderful timber that lasts forever underwater. When the tree had been sided, the six-inch board was marked up for the keel, stern post and deadwoods and cut out again with the pit saw.

Laying down the keel

The keel is the backbone of the boat and here it was 30 foot long, six inches wide and a foot deep. We cleaned it up with a plane. Then, on a slope down to the water, we set it up on blocks of hard knotty stuff, high enough from the ground to get keel bolts in.

Iffe had told me that his father Thomas had been a boatwright. He built skiffs for the smacks, gigs and jolly boats for the ships, all built under cover. Building in a shed you can hold everything down with a shore from under a beam in the roof: bending the planks round has a tendency to rise up amidships. He called this beam the "Bible" because you plumb all the measurements down from it. The boats had twelve apostle planks a side, which are fastened to the stem at the bow: "The stem that all follow, cutting the waves to salvation." Iffe said he was very religious and saw the practice of his craft as a kind of prayer. The vicar took a dim view of Tom's ideas, but Iffe said he could see the sense in it, as do I.

We were not building by measurements but by eye and rule of thumb: she will end up the size she ends up. Without the luxury of a shed and no overhead beam to shore down from, we had to rely on weighing down the planks as we went, then secured them with floor timbers athwartships.

I had to explain to young Will that fishing boats are clench-built, that is where each plank overlaps the plank below and they are fastened, or clenched together, with copper or iron nails clenched over a rove (like a washer). It is how the Norse men built their longships, rather than how larger ships are built, with smooth hulls like the Romans did.

So, first we positioned the keel on wooden blocks, level and straight. Then we cut out the stem and stern posts, both of which have inner timbers: at the bow this is called the apron, and at the stern it is called the inner stern post. All the upright timbers are held in place by knees we call deadwoods. The stern deadwood is longer and deeper to give her a fine run aft. All are secured together with clenched bolts. With a small boat the keel can be laid on its side to auger the holes and drive the bolts, then be all raised up together.

It was strange trying to find the words to best describe all this when so much is done by feel and eye, but I did my best to show young Will. Now he could see the boat taking shape he was beginning to get excited by the thought of sailing away to strange lands ruled by little people and giants and talking horses. I told him the tales of Gulliver are all made-up allegories. I then had to try and explain an allegory: "It is like seeing a reflection in the water. It shows you a different way of seeing something you are familiar with. You can see a ship as a body, with its keel being the backbone, its timber the ribs, and planks being the skin. Of course, *Gulliver's Travels* is also satirical, it's poking fun at authority." To which he gives me his stupid grin.

"All right, all right! Back to the job at hand." I was beginning to regret having read those wild adventures to him, but it had encouraged him with his reading.

Young William was so full of questions. I could see why old Iffe had told me to just watch him when I was an apprentice. I did not have the words to explain the shape of the boat I had in my head from my years of experience. I told him the best shape of a boat is like a fish: for her to have the head of a cod and the tail of a mackerel.

Planking is the skin

The first planks to go on were rebated into the side of the keel and went onto the side of the deadwoods, apron and inner sternnpost. These are called garboard planks. The garboards needed a bit of help at each end to get the twist on to the deadwoods, stem and stern post, so we held the plank over the fire for a while with wet rags around it. Some do not worry about charring the the plank, saying it seals the timber. The planks were held in place with gripes, which are like wooden tongs or cramps tightened with a wedge that nips the planks together where they land against each other. As we added the planks we used wooden thumb cleats fastened to the inside, to position the plank in place at one end while we bent it into position. When marked up it was removed, edges cut, and before putting it back we used the top edge to mark up the bottom edge of the next plank. The boy was getting the measure of it all and was soon marking up and cutting the planking on one side with me on the other. We helped each other when offering up, bending and fastening the plank.

Floors and frames are the ribs

We had already cut the floors, from the smaller upper branches of an oak tree, into four-inch boards using the pit saw. We cut them roughly to shape using a thin board as a template, then laid the timber into position and scribed, "spiling out" as we say, marking the exact shape for each side of the floor timber. I showed young Will how. "See here, Will," said I, "the widest gap? Find or cut a small block of wood to match the size of this gap. Then use that to mark the wood to be cut away. Make sure when you mark the plank's edge you mark it plumb upright. If you do not it will not fit, see?" He soon got the idea, marking and sawing down to the lines and chopping out with an adze.

The floors are the timbers that cross the keel athwartships. They need to reach as close to the "turn of the bilge" as possible. The turn of the bilge is the part of the boat that hits the ground when the boat lies on its side when it's not afloat, so for this reason this plank is thicker. It is important that all the floors are pegged in place with treenails* with wedges at each end, and that the first half-dozen planks are held down and secure before proceeding any further with the planking.

Next we made the curved frame for the widest the part of the boat. It was made up from boards we'd cut from the natural bends of a hedgerow oak; the twisted branches of the oak have

* *Treenails*, dry oak pegs (pronounced trunnels)

all the shapes you need for a boat. The pieces were staggered together like bricks in a wall to make the curve we call a futtock. It's important that the the grain of the wood follows the shape of the boat.

So, the first futtock was set up about a third of the way back from the stem (the widest part of the boat we call the beam). This futtock, frame or rib, or whatever you want to call it, was fitted to the planking that was already in place alongside the floor timber, and left oversized where the planking was yet to be. We bent a thin batten from the rebate between the apron and stem around the widest futtock to the stern post at the height of where I felt the top of the top plank would be. This is called the sheer line, and it's where your eye really comes in: there's a lot of adjusting until it looks right. We set up a futtock on either side between the first futtock and the stem and stern posts, pushing the batten out to a fair curve – that way you arrive at a shape that looks right. Between the sheer line and turn of the bilge, we marked the frames with the positions of the tops of the half-dozen planks still to be fitted. Then we used another batten to determine how much bevel needed cutting off the frames. The frames towards the bow needed the fore-edge trimming back, as did the aft-edge of the stern frames, until the battens sat flat on the surface of the frames. We adjusted the top batten by eye to show the sheer of the boat.

Some just carry on planking up without fitting these frames first, holding the planks out by shoring off the planking on the opposite side, and fit all the upper futtock frames once the planking is finished. Either way, it's good to have them in before bending the thick top sheer plank round.

Young Will. He wants to know "Why?" all the time. Some questions I just cannot answer. I say: "Just watch me! It takes faith to build a boat, Boy." If I'm honest, sometimes I'm not sure why, I just know when something is not right.

I almost forgot, a small half-transom was already let into the stern post. Above the small transom we let the uppermost batten go on past the stern post to represent the counter over the tuck. The top plank is thicker, called the sheer plank or wale, or gunwale. Then the rest of the futtock frames were fitted to the upper planking in the same way as the floors had been. With the help of a fire we bent the inner wale round the outside until it cooled, then fixed it inside, it being easier to bend a plank from one end than from its middle. The inner wale is what the deck beams sit on. Deck beams, deck hatches etc.

Launching

We named our little smack the *William and Mary*, a tuck-sterned sloop, she is thirty-two foot long, eleven foot wide and three foot three inches deep. Similar in shape to a ship's longboat but with a tuck stern and about a foot less in depth, less freeboard to help with the hauling of drudgers and nets and less draft for working in shallow waters. A tuck stern some call a lute stern is a small transom, then the rudder runs up through a small counter. The counter protects the rudder when tied up with other boats and gets the mainsheets a bit further aft. Young Will's eyes glaze over at this! So I say: "This is just the traditional shape that has been settled upon over the centuries. It works the best."

It was time for a change; time to practise some different skills. I had become too set in my ways with my understanding of materials and how to bend them to my will. But building this boat with my boy made me see again with the excitement of a young man's eye the possibilities of adventure. My adventures had all been within my art and this small river.

So now to rig her. Four poles made: mast, bowsprit, gaff and boom, all in the same way as I described earlier in my journal. We purchased all the rope we needed from Michael Harrison up at the ropeworks. We collected enough three-inch rope for the standing rigging. I could see young Will's eyes widen when I asked for three-inch rope, so I explained to him that rope is measured by circumference. Thus a three-inch rope is one inch across in diameter. We bought some two-inch for the mainsheet and throat, peak and jib halyards, and a little one-and-a-half-inch for the jib and foresail sheets. Harrison told us the hemp the rope was made from was from Riga but sometimes he gets hemp from St. Petersburg.

Cousin Thomas King was working over in Harwich for Joseph Munt, the blockmaker for the Royal Dockyard. We ordered fifteen blocks with elm shells and iron wood* (lignum vitae) sheaves** and nine elm deadeyes. We only needed four double blocks (with two sheaves) and eleven singles. I have heard tell that a large warship will need over a thousand blocks.

Thomas said they have machines in some of the dockyards to do all the donkey work. "Donkey work?" said I. He explained that the block mill, lathe and crosscut saw to work the lignum vitae would be powered by a pair of donkeys. Four generations of skilled blockmakers, and one day they will be replaced by machines and donkeys!***

Once we had all the spars and standing rigging set up, Charles Stacey came and measured up for the sails. The sailcloth was flax. It is the sails that bring the boat to life.

The boat finished, I walked along the quay with my boy to our new boat early one winter's morning, feeling the weak sunrise on my back. There was a faint smell of coal fires being lit in the crisp air, the mist that we call the daggs still hugging the river and Fingringhoe marsh with the bare distant treetops poking through far up the Roman river. A weak light from the mill. I could just see a group of dunlins, godwits and redshanks working the edge of the slowly advancing tide. It was one

---

* *Lignum vitae*, a very heavy and hard wood.

** A *sheave* is the wheel part within a pulley.

*** By 1799 Thomas' prophesy had come true with the adoption of a machine at Portsmouth Royal Dockyard. Ten men could do the work of over one hundred.

of those mornings that you felt the joy of being alive. It was tempered by the sad thought for those two foolish lads who had recently tried to cross the water bridge on a horse when there was too much water over it. The horse missed the edge and threw them off. One drowned and the other clung to a moored boat and was pulled out speechless. Water bridge* is what the local journal called it, but we know it as the graving bridge or ferry bridge, a low causeway that the water can flow under to cross the river from Anchor Hill at low water, without walking in the mud. It is prone to get slippery. The thought of losing young Will was too much. And besides he was turning into a good worker and had the makings of being a better shipwright than I ever was, if I did not make an oysterman of him first.

* *Water bridge*, *graving bridge* or *ferry bridge*, is a boarded causeway that follows the contour of the river bed (the grave) enabling you to cross the river at low water: it just keeps your feet out of the mud.

6
Oyster Dredging

The blacksmith made me up a couple of triangular drudge frames and I made a cow hide belly with long slits cut into it to let the mud escape, dressed with tallow and stitched onto the hoe blade we call the scythe. The blacksmith had punched holes along the back edge to attach the hide and a chapstick at the tail end. Next I made up a bit of net to form the basket that would hold the oysters and lashed that to the frame, the sides of the hide and a short chapstick. The chapstick is used to turn the catch out onto the deck.

I managed to get a licence to dredge with a little help from my dredgermen friends. Me and the boy had sailed with them a few times and learnt the ropes. Then we rented a couple of pits on west marsh and some layings up South Geedon creek from young John Sanford who had recently married Deborah, the widow of the oyster merchant Robert Hopkins. I remember young Will was friendly with his son Michael. I paid my share to the watchman who lived in the shed on stilts by the pits. I think Stanford pays his manorial rent to the Corsellis family with oysters. John and Deborah live in new Trinity House over the old shipwright's place, with his oyster pit and packing shed at the foot of Anchor Hill, just to the right of his grand bow windows.

The boat worked well. I could handle her on my own but to have young Will with me made everything easier. What joy to sail a fair south-westerly out of the river or to catch the last of the summer sea-breeze home.

We would sail to wherever we wanted to dredge and throw the dredges over the windward side, making the warp fast with a half-hitch on a thole pin in the rail. We were getting a feel to how the dredge was working. If we sailed too slow the warp would be slack. If we had too much sail up you would feel it bouncing along the bottom, so we would drop the gaff peak or trice up the mainsail tack, and back the jib (heave to). We moved the pins to get a balance – the aim was to be blown sideways by the tide. Any steering was done with the sails and the position of the pins along the rail. You wanted the pins to break if the drudge fouled, with the bitter end of the warp made fast for'ard outside all the rigging, or better still tie the warp to the scythe and just lash the warp to the ring, so if the drudge fouls the lashings break and the drudge upends and frees itself. All sorts of funny-looking rigs were used to control the speed, "scandalising" the sails. One dredge would be pulled, emptied on deck and thrown back, then you sorted your haul and the next dredge would be pulled, and that was how the day would go. Filling sacks with oyster or shell and five fingers and other pests like whelks and tingles* for the farmers to put on the land.

We drudged throughout the winter for broken shell from the sandbank where it naturally accumulates and put it on the marsh ready to use on the layings and in the pits. The new year started with raking the beds off of soft mud and getting the beds as flat as possible. This was done using a dredge rake that old John Stevens the blacksmith knocked up for me, and

* Oysters' main predators are *five fingers* (starfish), *whelks* and *tingles*.
Tingle is the local name for the oyster drill, also known as the *sting winkle*, that preys on young oyster spats.

finished off by hand when the tide was out, making sure we wore splatchers, boards tied to the soles of your seaboots to stop you sinking into the soft patches. The layings are marked with sapling poles tall enough to be seen at high water. We call these withys, bacons. After cleaning, on a neap and spring tide, we replenished the lower edge of the layings with stone and chalk. After the spring cleaning there was constant dredging to keep control of the pests, whelks, and five fingers or even mussels. Before the oysters spawned, the edges and middle of the layings were lined with broken shell. The pits were also cleaned out and prepared in the same way, any oysters that overwintered re-laid to fatten up and spawn.

Then we dredged the common free ground out in the estuary, re-laying them up the Geedons. As spring turned to summer we looked for signs of the oyster "spatfall" spawn. The water had to warm up. This was usually early June. The spat would attach itself to a stone or shell now the laying was left alone.

By mid-September with the pits full of sellable oyster we began to reap our rewards through the months with an "r" in them. The small ones were left in the pits to overwinter, you just prayed for the winter not being too hard and killing them all. It is the closest a fisherman hunter-gather comes to being a farmer! It was a tempting thought to carry a load of oysters over to Dunkirk and smuggle tea and brandy back but I thought better of it. Whitstable is the furthest I ever traded.

The first few years young Will would sail and work with me. But come his fourteenth year I could see there was no future for him in dredging. I asked old Iffe he would take him on as an apprentice. Iffe agreed. It was good for Will to get out of my shadow.

On fine days there was nothing better than being down the river away from everyone.

Some mornings working on the layings, up Geedon's creek,
with the boat heeled over in the mud,
mist still hanging on the marsh
a hazy sun that cast no shadows.
The popping of the mud and trickle
of the last of the ebb tumbling out of the creek,
The oysters squirting their disapproval into the air.
There deep in the dark grave of the creek
with hardly a sound
other than your own
And the muffled curlews cry
Light and sound paired back,
reduced to the heart of all.

But it was a very different matter in a foul east wind trying to beat out to the bank of shell with sleet and snow hitting your face like needles. Or trying to get home after your jib has blown out, torn to ribbons with a howling gale at your heels. The ropes and sails frozen stiff. Your face and hands swollen with the cold and cut to pieces with the handling of the crustaceans. I was told a trick once, that if when you first set out you plunge your hands into the icy water then hold them under your armpits, your hands will stay warm the rest of the day. It only works for a while. Boats! Better to build them than work them. Dredging is really a young man's game.

But the boat worked well and was much admired, so in the end I sold her to the John Tabor family who had moved down river to Bricklesea. And turned my hands back to what they were made for, building boats. I could see young Will had no real interest in taking over the dredging. He had the makings of being a fine shipwright, better than I ever was.

7
Wedding and Old Friends

Having finished his apprenticeship Will was now of the age to wed. He met Sarah in much the same way as I had met his mother, at church.

Come the day of their wedding I was bursting with pride. The couple looked so elegant.

Sarah had a white satin dress and looked wonderful. I had tried to persuade Will to wear my father's best woollen frockcoat which I had wore at my wedding but he was not having it. In truth it was a bit moth-eaten and looked very old-fashioned with its long lace collar. As we had just sold the boat we had money to have a fine new suit made by Page the tailor up river. They made a beautiful pale corded silk frock coat with a little embroidery round the hems and cloth-covered buttons, a high collar with only a little lace at the cuffs. The waistcoat was more heavily embroidered with breaches of plain corded silk. I had the necessary repairs made to the old frock coat and had it re-pressed by Will's tailor, and I wore my father's old clothes again. They will be good enough to bury me in!

The wedding was all very fine, much grander than my own. The memory of my wedding to Mary gave me a touch of sadness, but Captain Best had sent down his chamber musicians: two violins, a cello and a viola, the new combination. They lightened my mood and it was very enjoyable watching the young ones dance and dance. There's nothing like the voice of the cello and violin, I swear they played my very heart strings.

To my surprise old Clear turned up, as full of stories as ever. "Where is your parrot and foreign wife?" I asked, and he replied by asking if I had found that string of pearls. We laughed. I thought to myself, "My pearl died young," but did not want to darken the moment. Instead I told him: "My family are my string of pearls," and we raised our glasses to Will and Sarah.

He told me that after we last met he had a few more voyages back and forth across the Atlantic but his ship the *Rights of Man* was stopped by the naval blockade and he and some others of the crew were press-ganged into the Navy, some ships engaged in blockading the American coast. "I ended up serving in *HMS Mercury* a twenty-four-gun sloop under Captain Samuel Goodall. Got my wounded leg during the battle taking Havana from the Spaniards. The Navy made an honest man of me," he said with a smile and a twinkle. I remembered that look. He said he was no longer fit for the service and had returned to Wivenhoe to live with his maiden sister, "who has agreed to give me berth."

Old Iffe was also at the wedding. We laughed and greeted him. I was pleased to see he had no hard feelings towards Robert, "so long as you do not want the money back, the money your family paid me for your apprenticeship." That made as all laugh the more. Iffe was as keen as me to hear more of Robert's adventures. Robert told us of the wonderful shipbuilding timber grown on Bermuda, a dark red cedar that could reach heights of fifty foot and have a diameter of up to four foot. I have heard they are building very light, fast sloops with this timber rigged as topsail schooners.

After the wedding day young Will and I settled back into building more sloops, mostly for local fishing families. We even built a few alongside the house, dragging them across the marsh to launch them. New dredging licensing laws came in 1763 with a bailiff to police them. I would no longer have got away with not having had a seven-year dredgerman apprenticeship. But we had already given up on it. Now we worked for ourselves and a bit for George Wyatt. He did not have much going on when he first got going. He was fixing up a seven-year-old two-decker, a 150-ton snow from the West India trade, and selling a few old smacks, one with a Dutch wet well which had all her stowboard and trawl gear. Wyatt really began to get going in the yard with the building of the 62-ton *Mayflower*, a 54-foot, pink-sterned sloop launched in 1757, and the 90-ton ship *Providence*. More and more I felt I was a help for young Will with my knowledge, rather than my hands. He might say different!

Thomas Tunmer had remarried a Sudbury woman in 1762 a couple of years after Anna's death. He had given us our smallpox inoculations and was happy to attend the birth of Will and Sarah's children, the first one another Mary and the second they named Faith. He had done well for himself, looking after some rich and powerful families. He had studied midwifery since my losses and his own, and offered this service free to the poor. "It is appalling the number of deaths of mothers and babies that I have witnessed," he told me, giving me a look as he said this, and I nodded in agreement. But these were happier times. Young Will and I spent pensive hours waiting, pacing the quay, sharing a pipe or three.

It was also good to have Robert Clear back in the town. He joined in our joy of new life and for christening presents he gave little Mary a carved ivory monkey fist ball – he said it was from a narwhal's tusk – and to little Faith he gave a scrimshaw whale's tooth with his warship on one side and his Bermudan sloop on t'other. We would meet up in the pub a couple of times a week and swap yarns, he never seem to run out! We were very different in nature. He had never been married and as far as he knew did not have any children. From what I can see the Navy had been good for him, worked out some of his devil-may-care, or perhaps it was just old age. Our new friendship was one of mutual respect. I was very sad later when he died.

The last ship I saw Will working on for Wyatt was a 57-foot tuck-stern sloop. She was launched in April 1772 for John Oathwaite, a Colchester merchant who also had the *Providence*.

By which time I had become an old man too weak of limb to swing an adze or haul a drudge.

It has always been a pleasure to see the beautiful revenue cutters on the river. Captain Daniel Harvey became master of Martin's old revenue cutter *Princess Mary*, at that time the largest customs vessel in the country, 80 tons with a crew of four and twenty. This was after Robert Martin died in 1763. Harvey had been her captain for three years. After Martin's death Harvey made a new contract with the Customs and got more money per month per tonnage of his cutter. Daniel's grandfather John Harvey is said to have become the copyholder of Robert King's wood yard. His son, Daniel's father John Harvey is still alive. He was captain of the customs house cutter *Jean Baptiste*. And before that he was in the Lisbon fruit trade.

There are now faster and bigger revenue cutters working the river. You should see Harvey's fine new cutter *Repulse*, a more beautiful vessel never afore seen with a crew of eight and a boy. Then the second *Repulse* built two years later was 132 tons with eight carriage guns and six swivels and crew of thirty. Captain Daniel Harvey is doing well. Building himself a fine mansion here on the proceeds of all his privateering. State-sanctioned piracy! The rich and powerful sometimes get away and behave more like pirates, where the worker is shackled by their laws that the rich feel themselves above.

8
Ebb Tide

My life seems to have been punctuated by losses and gravestones. Looking back I see how these have shaped my life. In my early life these losses seemed to just drift past. It is only now the feelings catch up with me. I have had a long and contented life here by this tidal riverside town, a town I have grown to love. It seem ironic I have spent most of my time building boats, vessels that can transport you into other worlds with the temptation of wealth and adventure. But for me the boat also represents community, made of its many parts working together, that give a resilience and continuity within this turbulent world.

Let me describe the house in which I now sit. It is a fine proportioned brick-built tenement on the quay with a split front door. The room I sit in is to the right of the front door and chimney breast. We are on the end of the row of quay houses with windows on the side looking east down the river to Bricklesea church, over the small brook that runs from the mill at Bobbet's Hole and the ballast quarry. The ferryman's house is on the other end of our row, where at low tide the road crosses the river to Fingringhoe and where the ferry boat crosses when the tide is high. Carts cross the ford at low tide. The tide comes right up that road, cuts us off from the rest of the quay so we have a little cut behind our house to get round the water up to the Black Boy. There is also a road that runs up from the river to Love Lane, past the Sun. Something the town is not short of is ale houses.

This was my first home with my beloved Mary, the home I now share with Will and his Sarah and their young family.

Sitting here in this south-facing window with a knitted shawl about my shoulders, I watch the clouds play down the Roman River valley, while young Sarah busies herself adding all sorts to the pot on the fire, preparing a fine meal for us all for when Will gets home from the yard. The warmth reminds me of how things were here with my beloved, and of sitting at my grandmother's knee hearing family stories about all her long-gone relatives. There is not a day goes by when I do not think of my dear Mary in this very house. She would have just loved to have grown old with her grandchildren at her feet. I still share my thoughts with her. Will's youngest has the look of her with her fire-red hair and quick fearless wit. The oldest is perhaps more like me.

Nothing I like more these days than sitting watching the river's ebb and flow, the ever-changing water and clouds, the birds and the boats going about their business, the ferryman crossing back and forth. The continuity of life always different and always the same, the river ebbs past and floods in with new life. Future, past and present. The beginning and the end, all at the same time. Here I sit at this window reviewing my life. A whole and full life spent on and by this very river.

It is becoming harder to tell dreams from fact. The world is now more baffling to me with revolution in the America colonies. It is thought that a revenue man from down on the south coast added fuel to the fire when he wrote that America was European and should be allowed to trade freely with the rest of Europe. Thomas Paine's pamphlet was called *Common*

Sense, reminding us that in the eyes of God we are all equal. It makes a lot of sense to me, change is in the air. But I am not sure where it will all lead, it seems the old certainties are being swept away at such a pace. They have a prevention for smallpox, they say there has been made an ingenious beam engine pump powered by fire and water that can lift water and has the power of thirty horses. Whatever revolution or innovations take place human nature at its core remains the same.

If Bunyan has it right there are many different paths to Heaven. I have been blessed with a long life but I am now weary. If I have made sufficient atonement for my sins, by the grace of God I will talk again with my beloved Mary and see our poor lost children and my dear father. Perhaps I will meet my mother for the first time.

The nightingale still sung
as the blanket of night
gave way to cool dawn.

The full moon with
Faithful morning star
Shone in on William,
clammy and cold.

The light danced luminous,
on the dark river wave
Echoed, scudding night

Reminding Will
Of that elusive pearl
That danced
Wedding night bright.

All those tides ago
Beneath his beloved's
Wild free copper hair.

Last breath let go
sinks deep into
that still dark
warm room.

Part Two

The journal is now taken up by William son of William King

9
Father

Father died peacefully in his favourite Windsor chair at the window. I remembered his wish to be buried in his father's old frock coat, waistcoat, breeches and lace: the very clothes he was married in. I found them in his old sea chest along with this journal, a foreign-looking cutlass and some old papers. I helped Sarah wash and dress him ready for the burial. I hope that horrible woollen frock coat will get renewed before he meets my mother and his Maker. The churchyard is getting so full it is becoming a problem fitting more burials in, but Father got put in with Mother under a simple headstone with just their initials. My Sarah always grew a few sweet peas around the house and they were glorious that year. The day after the burial we cut a bunch and laid them by their headstone, and from then on each year this is something we like to do.

George Wyatt died the same year as Father, 1776. His impressive tomb and funeral was paid for with the sale of a smack called *Slowly*, as instructed in his will. I got to know George's apprentice William Stuttle when he started in 1757, which is when I started with Iffe. William married George and Mary's daughter, another Mary, in Oct 1764, same year as me and Sarah. William told me that he and William Popps, the new ropemaker, were brought in to witness George's will: he bequeathed his shipyard to his wife Mary and their dwelling house to his son Augustine, aged nineteen. We saw Mary Wyatt's announcement in the *Ipswich Journal*: "All persons having any claim or demand upon the Estate of George Wyatt

Shipbuilder... of this parish are desired to send an account to Mary Wyatt as she intends to carry on the business for the benefit of herself and family."

By 1778 Stuttle had established himself a new shipyard at the end of Grocer's Quay at the Hythe in Colchester. He and Mary were still living in one of John Sanford's houses in Wivenhoe until William built a new dwelling at his shipyard. The first boat William built there was a fishing smack called *Friendship* for Mr. Sanford.

Some time after the funeral I sat down and read Father's journal to Sarah. We noticed many similarities and a few differences in our lives. I've had the desire for some time to fill the empty pages. He lived his whole life on this river so was no Gulliver, but with his quiet interest in the little things he had a good understanding of human nature, and from his books the wider world. Now having a family of my own I have a better appreciation of the way he was. I am grateful for the skills he passed on to me. He did prefer his own company but to me he was always generous with his time, more like my older brother than a father. The old fool's only adventure was going oyster dredging. He did get a bit soft in the head towards the end with all his philosophising!

I have now reached the age that he was when he penned his story, so now I will have a go at adding my view into the next century.

10
Early Days

My childhood friend was Michael, the son of Robert Hopkins the oyster merchant. Robert died just before Michael was born leaving his mother Deborah with three daughters and a new baby, Michael. She married John Sanford with whom she had another six children. Michael was a bit older than me, very talkative, and for a brief summer we were inseparable. Together we explored the river in a bumkin, an old smack's skiff we named *Antelope*. We had a great time learning the ways of the water.

Father read me the first two books from *Gulliver's Travels* and when the chapman was in town selling books he bought me a children's illustrated edition with woodcuts to encourage me to read. He told me stories of an evening, tales he had been told when a boy. One was about a black dog with burning red eyes that stalked the marshes at night. He warned me never to go out on the marsh at night.

When I was twelve Father and I started building a boat. Michael would sometimes come along and lend us a hand but it did not hold the same excitement for him. He soon started working the oysters for his step-father. I think Father had always wanted to build a boat for himself. He said he had met a disreputable old friend in the Ship one evening who seemed to have had the most adventurous life at sea. Leastways Father said he told a good story. Made him think he should get on and have an adventure of his own. That and our inheritances from great-grandmother Mary and Grandfather Hall, pushed him to take a risk into actually building himself a sloop.

At first I had no idea how to go about building a boat. Father worked me though it stage by stage. At times he seemed not to have the words, so he would just show me instead. Showing me how to stand and how to hold and use each tool and how to sharpen them. How to read the wood, to know in which direction to saw, plane the wood or chop with an adze. He'd say: "Always chop into a knot". It was hard at first to understand what was in Father's head, sometimes he would get cross from all my questions. I was always wanting to know what we had to do next. "Patience. One thing at a time" he would say. As the boat grew, so did my understanding. This is how I learnt the shipwright's trade. In his way he was a good teacher.

But the old fool had the idea we would make our fortune with the oysters. It did not quite work out that way and was much harder work than building the boat had been. Looking back, my heart was not in it but I could see at first what pleasure it gave him to break free from the town. In truth, on a fine day there was no better thing than sailing out of the river, working the tides and harnessing the invisible power of the wind.

When he could see me feeling low he would say:

"The tide always turns when it reaches its lowest ebb". Or:

"The sail works at its best just afore it starts a-flapping. Ease your sheets!" Or:

"If a gale hits you, just let go. She will ride up into the wind. If you try and hold fast the gale will knock you down."

"The devil could've been a sailor if he had only looked up!"

We worked the oysters together until I was fourteen. Then I was apprenticed to John Iffe, Father's old master. I would have rather carried on working with Father but he thought it better for me to be apprenticed to Mr. Iffe. That way I could get my trade and he could follow his dream. "Besides, old Iffey could do with a strong young lad who knows the job," he said with a smile. I still sailed with Father whenever I could.

I got on well with Mr. Iffe. He told me that when my father started working for him, "he thought he knew better than I!" I now understood Father's smile.

Around this time Michael came to find me.

"Well William" he said. "Here we both are at the start of our working life."

He excitedly told me he was leaving on the morrow to sail with Captain William Hutchinson on the *Godolphin*, a three-decked East Indiaman with a crew of 99 and 30 guns, rounding the Cape of Good Hope bound for Madras and then on to Whompoa on the Chinese Pearl River. It was not uncommon for sailors to write a will before going to sea for the first time which is what he had just done, so he recited some of his flowery words to me:

"I Michael Hopkins of Wivenhoe in the County of Essex, seaman, being of sound and perfect mind and calling to mind the uncertainty of this transitory life as well as the danger of the seas and intending through God's permission to sail with the first fair wind to the East Indies..."

That was the last time I saw him.

One of Michael's shipmates from the *Godolphin* later came to Wivenhoe to find Mrs. Sanford with a note that he had taken from Michael and been entrusted to deliver to his mother. Deborah had kindly sent him on to find to me. He told me of their voyage, calling at Cape Town for water and

stores and a bit of recaulking, and then on to India and the East India Company's opium warehouse: "the largest in the world!" Then on to China to trade the opium for tea. Michael had died of fever on the return voyage. He was a good seaman and had helped this lad Dominic get over a fear of working aloft. I was pleased to know that Michael had a kind friend at his side on the voyage and at the end. He told us of the captain's fine words at his burial at sea. We shared our supper with the lad and Father gave him his bed for the night and a penny for his trouble. Dominic left next day on the packet to rejoin his ship in London.

I worked with John Iffe until I was one and twenty, by which time Father had given up on the dream of being a wealthy oysterman. So old William and young Will were working together again. We mostly repaired boats and built fishing sloops.

11
A New Family

I would often see Sarah Stacey in the church. I had in fact first met her when she was very young, when Father and I had collected the new sail for our *William and Mary*. Sarah would have been about nine or ten. She was looking after her younger sister in the corner of her father's sail loft. She caught my eye then and has since blossomed into a beautiful woman. Sarah's father's name was Charles and her mother was called Mary. Every other woman in this town seems to be called Mary!

Father has already written about the wedding but I would like to add that I never felt more of a king than on that day. The wedding and all, being measured up for my new suit of clothes... I had taken the tailor's advice on the combination of corded silk with a modern cutaway jacket and cloth-covered buttons. He did indeed make the most wonderful set of clothes. The tailor was George Henry Page at St. Leonard's at the Hythe. They made a lot of military and navy uniforms. While I was being measured up and fitted, seeing all the different shapes marked out with chalk that make up a jacket, I said to the tailor that it did not look too different to the parts of a boat. He said, "I hope my fabric is a bit more flexible than your wood!" We got a-talking and, seeing the tailor's son steaming and moulding into shape the breast panels of a bright red dress uniform, I said: "Again there are lots of similarities, you are making a curved shape out of flat materials. I use steam to bend a plank round a tricky shape." To which he replied: "Aye! That is something you can do with wool, unlike your corded silk."

The silk for the waistcoat was already embroidered round the leading edges before it was cut out. The tailor sat cross-legged on the cutting table in the window and tacked the parts together with big stitches. Then, after trying it out on me at the fitting, he did the final finer stitching. An apprentice boatwright I knew had a leg crushed by a boat; he took up tailoring and shirt-making. He said he felt it was not too different a trade. I have one of his shirts I keep for Sundays.

Sarah and I made a handsome couple dancing the night away to Captain Best's quartet. William and Mary Stuttle, who were also newly wed, were there. It felt like the whole town joined in the wedding celebrations. Father got a bit tearful after the brandy wine took hold and then there was the surprise of meeting his old friend Robert Clear again. I never met him afore but could tell Father was very pleased to see him. They were soon in deep conversation and when old Iffe joined them they got quite boisterous. By the end of the evening the three of them were arm in arm singing about some "Spanish ladies" like old Jack Tars. I'd never seen Father or Iffe drunk!

With our marriage settlement we were able to take over the adjoining tenement to our brick house on the quay from Grandmother Mary Hall. So we had double the space and no longer had to share the front door. That gave us plenty of room for Father and our future children.

Thankfully Sarah and I had survived the most recent outbreak of smallpox a couple of years before. Others in our parish were not so lucky. Sixty-eight souls in Wivenhoe were lost that year, mostly due to the disease. Father's friend Thomas Tunmer got to learn of the smallpox inoculations from Daniel

Sutton. I think they became good friends. Daniel Sutton made a fortune inoculating against smallpox after he successfully treated the inhabitants of Maldon in 1764. The stupid town tried to sue him for saving their lives! I think they have had a chip on their shoulders ever since they lost the battle of Maldon in the year 991. The courts threw it out and it helped publicise its success all the more. So Thomas continued this work in Wivenhoe. We were the first he inoculated. He said it was his wedding present to us. He inoculated Father at the same time.

Thomas told us that Daniel Sutton's wife Rachel died in Antigua when she was visiting her dying mother only a few years after their two children were born. Their son Daniel always claimed he was born at Wivenhoe Park, perhaps when they were visiting the Rebow family. Thomas had an impressive list of clients. He might have even delivered the boy. But knowing how the boy turned out I think Daniel made that up, as Thomas never mentioned it.

Young Daniel became a solicitor in Trinity Street, Colchester where he also became the town clerk. But he became famous for other reasons.

Daniel had a large boat store in Wivenhoe and a wharf where he kept his ship. Sanford always said he was a late payer. He called himself Vice Admiral of Essex, claiming certain rights over wrecks. But his appropriately named lugger *Fox* was really engaged in smuggling. The lugger was kept right under the noses of the customs on the wharf he called the Vice Admiral Quay! Anything salvaged in the Thames estuary had to be handed in to his warehouse and auctioned. Some would say he was a scallywag with a lawyer's swagger. He came charging up the river in his lugger with an enormous flag to announce Wellington's victory at Waterloo. After he sold the boat and stores he was helped by relations to emigrate to Tasmania.

It was not long before our first child was due. I felt excited and apprehensive. Sarah's mother was with us nearly every day as the due time approached. I was not sure what was expected of me so I just got my head down with work. At the first signs things were happening I sent word to get Thomas, he had said he would like to help with the birth. Things were well advanced by the time he got to us. Sarah's mother, as far as I could tell, had everything in hand. Father and me were shown out of the house. It was a fine bright morning with the mist just lifting from the water. The river looked like it was steaming. At the first cries I rushed in to find we had a fine dark-haired little girl! Sarah looked hot with pink cheeks, sweaty and tired but with a big smile from ear to ear. Thomas pronounced mother and child were in good health.

When all was settled, Father and I took Thomas up to the Black Boy for a glass or three of Dutch courage to celebrate a new King. I felt stunned like I was walking on air. The landlord just took a look at me and instantly raised a cheer, my smile said it all! We had already decided if she was a girl we would call her after her grandmothers, Mary. So we raised our glasses to Mary and Sarah.

It only seems a blink of an eye before young Faith was born. She had fair hair, green eyes and pale skin. As she grew her hair became red. Their baptisms at St. Mary's were further opportunities for me to wear my fine silk suit and pretend I was a gentleman for the day.

The girls would beg me to tell them the story of the handsome highwayman Thomas Coker, which went something like this:

"It was a moonless stormy night. Coker worked the Harwich Road, preying on weary travellers on their way to take ship or returning to England. There was a coach and four with a wealthy merchant and his daughter, late for the packet ship. Coker had noticed them at the inn when they had changed horses, and had ridden ahead. He waited at a dark and muddy part of the road where he knew the coach would be slow. Out from under the trees he rode on his fine black steed, pistol in hand, right alongside the coachmen, roaring those infamous words 'Stand and deliver! Your money or your life!'" This would always make the girls squeal.

"He was too quick for the coachmen. By now he had a pistol in each hand and another loaded pair in his belt. He ordered them to step down from the carriage, but the old merchant levelled a shot, wide of the mark, and Coker fired back and caught the poor old man in the shoulder. His daughter screamed and fainted away. The coachmen fell to look after the old man, only after Coker had relieved them of their valuables, then Coker revived the girl with some smelling salts he just happened to have with him. At which she promptly fainted away again, awaking in the arms of our handsome highwayman kissing her! He relieved her of her earrings and rode off into the night."

I warned the girls not to be taken in by any handsome men, be they highwaymen or not. Coker was caught in Wivenhoe boarding a ship for the continent. He was taken to Colchester Castle. It turned out he had many sweethearts in the inns of the Tendring Hundred. He had spent all his ill-gotten gains on drink, wooing and promises of marriage. The gaoler said he had several poor wanton weeping women visiting him in gaol with child, wanting to know where his loot was hidden before he was hanged. One had a particularly fine earring with which she tried to bribe the gaoler to let Thomas escape.

Mary also liked a story I made up about the fairies that lived under our floorboards. The fairies loved pineapple and making magic coconut pyramids. The king and queen would get upset when she slammed doors for it would ruin their baking with all the dust.

The girls were good readers and soon outgrew their chapbooks from the chapman. Sarah would read her books to them, which as far as I could see were mostly about a silly mother of some opinionated young girls trying to find advantageous husbands. I told them not to hurry into being married. Not that they listened to me on such things.

When she was seventeen Mary married Daniel Chapman the baker's boy, with an oven big enough to bake seven to eight bushels of flour* at a time. They did well, making ship's biscuits and the occasional coconut pyramid for me. Faith married Thomas Cole, a fisherman, who would often give us a half a bucketful of sprats in season. This made a good supper, dipped in flour and fried and eaten with a slice of fresh buttered bread. Faith and Tom live in the other half of our house on the quay with their two girls Elizabeth and Mary and two sons Thomas and William.

They are good girls. I bless every day that I have my family about me.

---

* A *bushel* is 8 gallons or 60lb. 8 bushels would be 480lb.

12
Building Ships

Mary Wyatt ran the shipyard for three years after the death of her husband George, but then sub-leased the yard to Moses Game. He got the contracts to building two large ships for the Royal Navy. I started working for him and we built them on Wyatt's old slipway up river from the Ship at Launch. I remember Father telling me the best slope for a slipway was five-eighths of an inch to a foot. That gives the right slope for launching. While talking about the slipway Moses pointed out to me the crab winch on the marsh opposite, "to get the ship started come launching day. With enough time and men, pulleys, levers and rollers you can move anything. That's how they built the pyramids, one stone at a time."

Father had given me his tools. Like the old boy himself most were quite worn out. I had to replace a plane blade or two, they were sharpened almost away, and his hand saw was down to half its original length! I had tools of my own, there are now some fine crucible steel tools available from Sheffield, but it felt good to have his tools and a few from his father's afore him.

The first of Moses' ships for the Navy was to be called *Inspector*, 97 foot long by 27 foot wide, a 310-ton sixteen-gun ship. Moses showed me the drawing the Admiralty had sent down. It was immaculate in every detail. Even her internal workings were shown in red pen. Keel and deadwood scaphs, iron work, even the carved stern and figurehead all drawn to scale. A thing of such beauty as like as I had never see afore. Moses himself took charge of scaling this drawing up to life size, he called it lofting out, and the making of the template moulds for us to mark out and cut the timber from.

A ship's construction is different from the small clench-built fishing sloops and smacks. The frames go up before the planking and the planks make a smooth hull, we call this carvel-built. The frames are all bolted together with iron bolts and washers like a rivet, and then set up, whereas in a clenched boat, the planking makes the shape first and the frames are fitted to them.

We set up the blocks over the slipway, constructing the launching cradle later under the ship. First was laid the keel, stem, stern post, and deadwoods, then all the floor frames athwartships and then the keelson over the lot, all secured and clenched up. Then we started to raise the futtock frames, working from the midship frame to each end. We put the whole futtock together on the ground then raised it into position with blocks and tackle from an overhead cable between sheer legs*. The A-frame or sheer legs were at each end of the ship and had been used to raise the stem and stern post, and a cable was rigged between them that was used to raise the futtock frames.

All the frames and transom frames &c were securely held true and in place with battens and shores, and cross spans from one side to t'other like temporary deck beams. Battens were used to see where the frames needed trimming to make them fair with each other.

Then we started planking up. Planking was three inches thick bent into place from one end. Temporary forelock ring

---

* *Sheer legs*: two poles tied at the top and set up like a capital "A".

bolts* were used; an oak wrain stave** was lashed to the ring of the bolt either side of the plank, under which wedges could be driven to bend the plank into place against the frames. A temporary flat sett was below or above the plank with which the plank could be set against the adjoining plank. The planks were steamed in a kiln to make them more supple where necessary for an hour per inch of thickness, plus one for luck. It would take at least a half-dozen men to manhandle the hot plank into place, fastening with one-inch diameter dry oak pegs called treenails, secured with a wedge at each end and clench bolts. The blacksmith made the bolt with a head on one end, and once driven though the augered hole the other end was hammered over a washer, like a clenched fist. Sometimes if the hole was a bit small the bolt would be driven hot from the forge to char the inside of the hole against rot. It would always make flames shoot out of the hole as it was driven but it made clenching up easer and as the bolt cooled the metal would contract and tighten still more. Moses did not like us doing this, afeared the whole ship would go up in smoke, so we had to keep dowsing the wood with water, steaming the planks where necessary, inside and out.

Then it was the caulking. The boys would be given the job of rolling the oakum on a sack over their laps, supplying the caulkers with spun oakum** of an even thinkness which they

* A *forelock bolt* has a slot for a curved iron wedge to tighten it.
** A *wrain stave* or *staff* is a short piece of wood, tapered at the ends.
*** Prisoners were often given the job of unpicking old hemp rope and gunny sacks of hemp, jute or sisal to make oakum. The oakum is then drenched in pine tar.

drove into the seams. Once the seam was all tight, hot pitch was painted over and scraped off the plank surface while still hot. Under the ship this had to be done several times to fill the seam. Her bottom was then copper-sheathed with hot tar, with cow hair between the copper and the planking. But on deck the hot pitch was simply poured into the seam from a ladle full of pitch, kept hot by a heated-up cannonball on a rod we call a "logger-heat".

The *Inspector* was launched in 1782. That was a fine day with the whole town coming down to watch her splash into the water. The job would only be half done come launching, as there would be the fitting out, ballasting and rigging. But the launching was the best day. There was much drinking and celebration that day. The Ship at Launch, The Anchor, the Bull and Red Lion up Anchor Hill, The Swan, The Rose and Crown, The Black Boy and the Sun all did great trade that day. Public houses is something we are not short of. Someone later told me they thought the *Inspector* had been decommissioned and went a-whaling in the Pacific, joining the British Southern Whale Fishery and was then captured off the coast of Chile.

The second ship we built was even larger, 109 foot long and 30 foot wide. We had her finished the following year. She was called the *Comet*, a 424-ton fourteen-gun fireship. Strange to build a ship when you know its function is to be sailed into an enemy ship ablaze. She was one of only two fireships that was actually used in that way in action during the French war.

It was good to work on ships of this size. It was work that did not come round that often. We were joined by some of the lads from over the river. I got talking to one of them, Thomas Clarke, whose father William was a master shipwright. He could neither read nor write but was very successful. William had just died and signed his will with an X. He left Thomas a couple of ships and an old river lighter which was very useful to work from when working on a ship afloat or getting materials across the river.

The Admiralty can be a bit slow making the final payments on new ships. Moses did everything he could to keep trading. I swear he would have parted the waters of the Colne if he could have! He sold the timber from a yawl he had broken up, and in the same year he launched the 70-ton ship called *Edward* but was declared bankrupt in 1783. Me and Thomas Clarke had almost finished framing up a cod smack over the river for him at the time of his bankruptcy. Moses was good enough to make sure we got our pay. Later on he built himself a seven-ton smack over there called *Liberty* and turned his hand to a bit of smuggling to try and keep his head above water. But she was seized by customs in Harwich and cut in pieces.

After Moses I was grateful to get work with my old friend William Stuttle up at the Hythe. William had built himself a fine new shipyard and house on the Wivenhoe side of the river at the end of Grocer's Quay, up above the river lock gates, so the water was always high, and his new graving dock also had to have gates. To drain the dock the water ran into a ditch behind the sea wall, flowed down below the lock and then drained though a sluice back into the river at low tide. William had lots of apprentices to manage. While I was there we built five smacks and three brigantines, the longest being the 66-foot *Concord* for Thomas Wall and a group of shareholders that included a tallow chandler and several grocers.

The graving dock was the very dock where Sainty was to later build the famous *Pearl*.

Philip Sainty had learned his trade apprenticed to William Stuttle afore taking over the Bricklesea shipyard.

13
The Walk to Colchester

On a good day it was all very fine walking to William's shipyard with the early morning sun on my back. I'd stop for a moment and look back at Wivenhoe though hazy morning light, the mist that the sun was yet to burn off. The river would shine bright as it snaked around the marsh and oyster pits, Wivenhoe church aglow against the distant wooded hills. The whole scene bleached out, or cut back by a cool river mist. Sometimes you could just see the upper parts of trees on each bank and parts of the town poking through the mist with the smoke from the newly-made-up fires hanging motionless above.

Turning back to the Hythe I'd see the red-tanned squaresails of a river lighter or small river boat as they caught the glow of the rising sun. In summer the anxious skylarks would hover high above the flat meadows, thick with buttercups, that ran down to the river. Some days the pale-green sea purslane vibrated with mauve sea lavender, and where the fresh-green shoots of samphire fringed the marsh, a patient heron would sometimes stand at the water's edge. The very mud seemed to have a purple sheen. In those moments I understood what my father was talking about when he said: "See all the colours in the mud!"

The Romans are said to have straightened this stretch of the river to help their ships to the first Britannic capital. The navigation of the upper part of the river must have been a problem even then. I have heard talk of putting a lock gate further down river just above Rowhedge and turning this part of the river into a canal with a basin somewhere behind Stuttle's shipyard. I think Hawkins' boy later tried to make this happen but made do with building a dock fed by Bourne stream below his new house. Which in the end I think got used as a mast pond.

Returning home again with the evening sun behind me, the whole landscape would be illuminated before me, the colours more vibrant and the shadows deeper from the low sun.

In mid-winter in driving rain or snow it was less pleasant, making the bright days all the more special. Bad days I would sometimes take the road and try and catch a ride with a passing farm cart. If only I could have afforded a stable and a pony.

Years later when I had given up working on a fine day I would sometimes walk up river to try and recapture the vibrance of those days. On such a day near where the park land runs down to the river I chanced to fall in with a young artist. He was staying with Major Rebow up at the big house. He talked about the clouds and showed me a dark convex mirror he called a "Claude glass" which opened like an oyster. It helped him to see the clouds and reduce the landscape to tones. By turning his back to the view and looking in this small dark mirror he could even study the setting sun. By this time my eyes were dim and I could not make it out, but it was good talking to him; he seemed to be very interested in the life of the working man, being the son of a miller up on the Suffolk border, he told me. With the money from the painting of Wivenhoe Park he was hoping he would soon be wed.

14

Revenue Cutter Captains

In September 1777 we read in the *Ipswich Journal* that Captain Daniel Harvey's grandfather, John Harvey, had died age 94. That's a fine old age. Father had died the year before aged 64, he had always liked to hear news of the revenue cutters' exploits. I can still hear him say "The *Repulse*! No finer craft on the river."

Captain John Harvey had been a revenue man commanding the *Jean Baptiste*. His son Daniel was also a revenue man, having commanded the privateer *Princess Mary*, a cutter with eight carriage guns and six swivel guns. In 1741 under the command of Robert Martin, the *Princess Mary* had captured a Spanish ship; the prisoners were paraded through Wivenhoe and taken to Colchester castle. Daniel had taken command of the cutter by 1756. She had a crew of 24. His gunner was Martin Hopkins, another Wivenhoe man.

In 1760 Harvey took command of *Repulse* with Hopkins as his mate. However by 1776 Martin Hopkins was in command. We were surprised to see in the *Ipswich Journal* in February 1777 that he was selling the old *Repulse*. We had no idea he owned her; he had done well. I guess he needed to, having ten children.

A QUANTITY of HAY. Enquire of Mr. Fancett, Crofs-keys-ftreet, Ipfwich. ☞ To be delivered in Ipfwich or Woodbridge.

To be SOLD by PRIVATE CONTRACT,
THE REPULSE CUTTER, with her Materials, lately employed in the fervice of the cuftoms, at the port of Colchefter. ☞ Inquire of Martin Hopkins at Wivenhoe.

THE Grounds and Premifes belonging to the WHITE HOUSE near IPSWICH having been, for fome time, notorioufly trefpaffed upon, to the great damage

His luck did not last. Three years later, Hopkins having been commanded to stay on station off the French coast, the *Repulse* ran aground and was captured. Hopkins and his crew were imprisoned for thirteen months until a ransom was paid. The *Repulse* became a French privateer until she was recaptured and repossessed by Harvey, who re-employed Hopkins as mate.

In 1782 Hopkins was back in command but was caught selling the seized goods back to the smugglers, and was dismissed. Harvey took command again, until he was himself dismissed; he had fallen from favour with his lordships when he was discovered to be claiming money for more crew than he had aboard his revenue cutter.

When Daniel Harvey died in 1794 the sale at the Falcon Inn of the contents of his grand house made interesting reading: so much furniture, including two large Turkey carpets and fifteen iron-bound hoghead casks. He bequeathed to Hopkins, late mate of the *Repulse,* an annuity of ten pounds per annum during his life, "as a testimony of my approbation of his fidelity and good conduct in office from which he has been most rudely and most unjustly dismissed." He probably thought the same about himself!

All these captains sailed very close to the wind with the law. I can just hear old Horace saying: "Those revenue men are nothing but pirates, out to line their own pockets!" Daniel made enough money to build himself a fine mansion up the road.

Captain George Munnings from Thorpe-le-Soken took command of *Repulse* when Harvey was dismissed in 1789 and wrecked her later that very same year off Harwich. He managed to salvage the mast and spars. So a fourth *Repulse* was built to replace her. This one was owned by the revenue service rather than hired. She was the largest of the revenue cutters to be based in Wivenhoe at 210 tons, armed with sixteen carriage and twelve swivel guns and a crew of fifty at a cost of £1,552-16s-8d.

In 1797 there was great excitement in town about the French privateer that was captured and brought into Wivenhoe by Captain Munnings of the fourth *Repulse*. The French lugger privateer was called *La Tigre*. We all witnessed it, but I read about it later in the *Ipswich Journal* which I cut out and saved. It was reported: "*La Tigre*, commanded by Capt. Chataing, having 29 men on board, carrying two 2 pounders and 4 swivels, with a number of small arms and cutlasses. She was chased above 3 hours, and taken within 3 leagues of Dunkirk. The prisoners were on Wednesday escorted from Wivenhoe under guard of the Northumberland regt. of militia to gaol at Colchester, where they are to remain 'til further orders." I do not think I had knowingly seen a French man afore!

A poster was put up for an auction at the Rose and Crown on Thursday 12th October for "*La Tigre*, prize to His Majesty's revenue cutter *Repulse* with all her arms and ammunition as now lies at Wivenhoe Quay. A deposit of 25 per cent will be required, and the vessel &c to be taken away with all faults, as she now lies, within seven days after sale, or the deposit to be forfeited."

Another success of Captain Munnings and the *Repulse* was the seizing of over three hundred half-ankers* of brandy and genever. By my reckoning that is over 1,200 gallons of spirit that all got locked away in Colchester customs warehouse. I am not sure how much of it got to be auctioned.

Captain Munnings was very hot-headed and not beyond throwing a punch in a disagreement. He was also not popular with our local fishermen, continually harassing them and even seizing their boats if he found they had infringed any of the regulations. He seized John Tabor's old smack *John & Sarah* off the Naze for smuggling in 1800. It was just as well that he lived over at Thorpe-Le-Soken. He also owned close to his home, probably the most secluded public house in Essex, the Kings Arms at Landermere Quay at the head of the Backwaters, just the right place for landing a bit of contraband of his own. He resigned as captain of the *Repulse* and went prize-taking in his privateer *Courier*.

* A *half-anker* held 3¼ gallons.

15
The Stacey Brothers

In June 1801 the yawl *Queen Charlotte* and the lugger *Fly** were auctioned by His Majesty's Customs on behalf of the *Repulse*. "No One Pound Notes will be taken, except those of the Bank of England; nor any Country Banker's Notes, except those of Colchester." Captain Daniel Stacey was the *Repulse's* commander by this time after Munnings went privateering. Daniel's brother Benjamin and his wife Sarah started off in Thorpe-le-Soken. My Sarah is a Stacey. I never worked out how she was related to Daniel and Benjamin. She just calls the Stacey brothers her young cousins. She told me the strangest thing: three of Benjamin's children were baptised in Thorpe-le-Soken between 1789 to 1793, and then all baptised again at the same time in Wivenhoe in 1805.

The Staceys built themselves a fine house and granary on Wivenhoe quay at the corner of Rose Lane, opposite the Rose and Crown just up the quay from where we live.

Daniel looked formidable in his leather sea boots, a cutlass at his side and a two pairs of pistols in his belt. But standing aboard one of the fastest ships on the coast with sixteen carriage guns, twelve swivel guns and fifty men at his command would be enough to strike fear into the heart of any smugglers or foreign privateer. I heard he chased a heavily-manned privateer lugger off the coast of Lowestoft. In 1802 in Wivenhoe he sold off his

* *Fly* had been cut into pieces, along with 3 other boats. It's normal practice to cut a vessel into pieces when they'd been caught smuggling. Sometimes these boats would be put back together, or the wood used for other purposes.

share of a 64-ton sloop *Thomas and Hannah*, in "Mr Stacey's Rose and Crown". Again, four years later an almost new 47-ton fishing cutter *Mary* was sold by him.

Daniel's first wife in Mistley was widow Elizabeth Swift. After she died he married Sarah. They had a daughter they named Elizabeth Sarah. He died a few years after, leaving Benjamin his share in three smacks and a ship: *Friendship Increases*, *Elizabeth Sarah*, *Welcome Messenger* and the snow *Venus*. All but the *Welcome Messenger* I had a hand in building, but more about that later. Benjamin also inherited his brother's pistols and sea boots. The quadrant and another pair of pistols went to his young nephew John. Daniel's household goods were all auctioned on 29th March 1808. I kept the advertisement from the *Ipswich Journal*. We would have liked a mahogany chest of drawers and a Pembroke table but were outbid.

Elegant, Modern and useful Houshold Furniture,
WIVENHOE, ESSEX.
To be SOLD by AUCTION
By BUNNELL and JACKSON
On Tuesday the 29th day of March instant,

ALL the Genuine Houshold Furniture, China, Glass, &c. of the late Captain DANIEL STACEY, dec. at Wivenhoe; comprising 3 sets of mahogany chairs, black seating, 2 arm ditto, 8 Windsor ditto, mahogany night ditto, mahogany card, dressing, Pembroke, and dining tables, (fine wood) mahogany double chest with drawers, handsome pier and dressing glasses, Scotch carpets and Wilton ditto, each 13 by 16, handsome sofa, 6 feet 4; good as new, with cotton covering and 2 pillows, a beautiful mahogany 4-post bedstead, with elegant chintz furniture, tastefully made up, with an excellent bordered goose featherbed and beding, several other good 4-post and tent bedsteads, various hangings, 5 very good featherbeds, blankets, quilts, and counterpanes, mahogany cheese waggon, butler's tray, waiter and tea boards, handsome tea and coffee urns; some parcels of very good china; 5 china bowls, 3 copper boilers, saucepans, with a variety of kitchen articles and other useful goods; all of which will be expressed in catalogues, to be had at all the principal Inns; place of sale, and of the Auctioneers, Colchester.

16

The Wivenhoe Yard: Hawkins & Sainty

Not long into the new century, after fifteen years of walking to work at Stuttle's or sometimes taking a boat if the tides were right, I was pleased to hear of a new shipbuilder, a William Hawkins, who was looking for shipwrights in Moses' old Wivenhoe yard. So after talking to Hawkins I thanked William Stuttle and went back to working in Wivenhoe.

The shipyard had passed via Wyatt's widow and son to Samuel Cook of East Donyland* over the river, who had subleased the yard to his nephew Joseph Cole, the Sainty brothers and Hawkins, the yard being subdivided.

Joseph Cole built at least eleven smacks in Wivenhoe but had also built several smacks before in East Donyland. Captain Munnings had a 36-ton smack built by Joseph Cole there in 1791 which he proudly called *Repulse* after his revenue cutter. She was 39 foot and six inches long.

Joseph's last Wivenhoe-built smack was the 46-foot *Joseph* which was seized for smuggling. Joseph died in 1815, and his son Daniel, who had worked alongside him, carried on building after his father's death. Half a dozen other people built a smack or two in the yard around this time.

Robert Sainty built four small fishing boats in the Wivenhoe yard. Young Thomas Harvey also built one smack.

I built at least nine smacks with William Hawkins between 1802 and 1806, including two 42-ton smacks for Daniel and

* *East Donyland* is now called Rowhedge.

Benjamin Stacey. *Friendship Increases* was the first one. The second was called *Elizabeth & Sarah* named after Daniel's two wives (or perhaps his daughter). A third share of the latter was owned by William Fisher who ran Daniel's London shipping office.

We also built two brigantines. The first of these was the *Venus* in 1803, for which Hawkins asked me if I would carve the figurehead. I had never attempted a full-size figurehead afore let alone a half-naked goddess. But I agreed to have a go. Going through my tools I realised my gouges were in a sorry state; I needed a really broad fishtail gouge. I had heard the best edge toolmaker was Mr. Addis in Deptford. I wrote to him telling him that I required a few fishtailed gouges of different sizes and would they write back when they were made, then I would call on them when the first available passage could be arranged. Fortunately there are many London-bound boats from Wivenhoe. I took passage with one of Sanford's boats delivering oysters to Billingsgate. They agreed to drop me off at Deptford and pick me up on the way back. It was good to be on an oyster sloop again; it reminded me of the days I spent down river with my father.

Once in Deptford I soon found the Addis workshop in Church Street. Joseph Addis told me they make a lot of tools for the Royal Dockyards, including for the figurehead carvers, and that they favour the widest flat gouge to work the first blocking out with. He told me his father Samuel had supplied tools to the finest of all wood carvers, the Gibbons family, who started in the Deptford ship-carving trade before being taken up by the aristocracy to embellish their stately homes.

I purchased the tools I needed. They were not cheap but of the finest quality of cast steel. I had some time to spare before boarding the boat home. Deptford is an interesting place; they've been building ships here for centuries. They say the Tsar of all Russia came and worked here to learn the art of shipbuilding. I blagged my way through the gates of the Royal Dockyard saying I was a shipwright looking for work. It is a fine establishment with grand houses for the officers' families, a clock-house giving cover to the saw pits, and a graving dock long enough to build two warships at the same time. It is extremely well-organised. Deptford Royal Dockyard is also from where Wivenhoe's famous *Nonsuch* set out for Canada on her first voyage of what became the Hudson Bay Company.

Inside the Master Shipwright's office hung a fine painting of a warship being launched in this very yard, painted by another artisan who had worked here by the name of John Cleveley. I could see he understood how a ship was put together. He had been a ship's carpenter and also an accomplished painter. The clerk told me his sons still worked in the dockyard and also painted. The Master was too busy to see me, which was just as well because I did not really want a job. And besides I was looking forward to getting home and starting work on the figurehead.

We took the last ebb as far as we could and dropped anchor just below Gravesend for the night. At first light just after high water we set sail again, and with a fair north-westerly wind and the tide under us we made good progress following the Essex coast round, between the Maplin and Barrow Sands in the West Swin channel. We then felt our way with the lead line over into the East Swin. Now out of sight of land we found our way though the narrow spitway between the Buxey and Gunfleet Sands before the next floodtide bore us up the Colne to home.

The blockmaker had turned me up a small round mallet head of lignum vitae, and an ash handle, and I was ready to start work. In 1700 the Admiralty had decreed that figureheads should no longer made from made of oak, elm, mahogany and teak as they were deemed too heavy; in future figureheads had to be made from lighter pine. So I first built a rough shape of the figure out of a good quality pine, pegging the arms to the body and cutting the mortice to sit astride the fore knee beneath the bowsprit. Then, with my fine new gouge with its eight-sided tapered handle and mallet, I started removing wood to reveal my goddess. Once finished to my satisfaction the painters took over, their flesh colours and ruby lips and nipples bringing the whole figure to life. Though I think I preferred her as I had left her, a more subtle beauty.

The *Venus* was for Richard Mills of Colchester and William Cowper of London, and was sold to John Mann (of the Colchester lighterman family), re-registered as a snow (with a higher tonnage!) and sold the very next day to Hawkins' son in Colchester and then in 1805 to Daniel and Benjamin Stacey.

One of my last jobs for Hawkins was the *Fox*, a square-sterned Brigantine, 94 tons and 81 foot long, launched in 1806. William Hawkins died in 1812. His son, another William, set about making a fortune expanding the timber business up the river.

Philip John Sainty took over the Wivenhoe yard from Hawkins. Philip's yard in Bricklesea had been declared bankrupt in 1798. Like Moses Game when faced by bankruptcy Sainty had turned to smuggling to regain some capital. Philip made several successful trips to Ostend in his cutter *Ruswarp* but he decided he no longer wanted to continue this work for the Colchester publican James Forster. Forster decided to turn King's evidence to save his own skin and ratted on Philip. The *Ruswarp* was seized and Philip and his crew were gaoled in Chelmsford in 1816 with a penalty of £4,426 unpaid cargo duty on their heads. His brother Robert was perhaps more the smuggler. Robert got chased by two revenue cutters down-channel in his lugger *Wolverine*. To get away, with them not knowing what his cargo was, he ran her ashore at Worthing and set fire to her.

Luckily for the brothers the Marquess of Anglesey, Lord Paget, had heard of their fast vessels. He paid the penalty to get Robert and Philip and their crews out of gaol on condition they built him a fast cutter. Major General Lord Paget had done well out of fighting Napoleon, capturing the French wagon train of Napoleon's brother King Joseph Bonaparte's treasure in 1813. Apart from losing his leg at Waterloo while he was sitting on his horse next to Wellington. To replace Lord Paget's love of riding and horse racing he had decided that racing yachts would be the thing.

After three years in gaol, Philip built the Marquess his yacht *Pearl* up in Stuttle's old yard, 92 foot long, 21 foot wide and nine foot deep. She is clench built with carvel wales, much the same construction as our fishing boats but a lot deeper at nine foot in the water, giving her a better grip to windward, a far better deep-sea boat like the revenue cutters. She is captained by William Ham, a Wivenhoe man. She was originally rigged as a yawl with two masts but has quickly been converted to a cutter. She was launched last year in 1820, up at the Hythe.

Sainty was paid a retainer of £100 per year by the Marquess not to build yachts for any competitors, but we all know he did not honour this agreement. Sainty built more yachts at Stuttle's and then in Wivenhoe; there was one called *Anglesey* for Benjamin, John Sanford's boy.

Sainty by name but not by nature, Philip had many wives, including two at the same time. And many many children and more than a few bankruptcies. The *Pearl* emerged from a rough old native oyster! It took a flamboyant lord with a soldier's loot linking up with a shipwright-smuggler to establish Wivenhoe's shipbuilding reputation throughout the land.

17
Fireside

We recently heard of a new church they call the Church of the New Jerusalem which started in Bricklesea a few years back. Meetings are now being held in Wivenhoe and we have started attending for our Sunday worship. We meet outside in Sun Yard, behind our house. It is good to hear the guiding words of the Bible direct from a fellow working man.

Since I stopped working I've been writing this journal, which reminded me of the papers in Father's sea chest. The other book turned out to be Captain John King's log for the ketch *Robert & John* of Wivenhoe. I wondered, was she named for Robert & John Page her builders or for Robert & John King her owners? From the journal I realised that the sword must have been Captain John King's. I have a distant memory of a tale Father used to tell me about young John's escape from privateers with a Spanish cutlass. This the very same cutlass we now hang over our fireplace.

The five sheets of loose papers were dated 1670 and seemed to be a copy of John King's will. It was hard to read but I could make out that John and Mary King's two sons Francis and William inherited an eighth share each of two ketches, the first the *Robert & John* of Wivenhoe, one hundred tons or thereabouts, and the second, the name is hard to make out, but I think it is *Batchelor*. James Lock was her master. Lock was married to John's daughter Mary. James and Mary lived in Marsh House. The youngest sister Susan was left property and an income.

Also bequeathed were various properties including the land called Burrs, barns, a wharf and a house that John built and dwelled in, all left to his sons and heirs in perpetuity.

"Well" said I. "That is a thing! The shipwright's place was on Burrs. The shipyard has been ours all the time! And the house Father grew up in was also down West Street adjoining the shipyard. I wonder why Father did not mention any of this."

Sarah thought about this, laughed and said: "That will is over a hundred years old! Your father's father bequeathed it to Aunt Mary. If the rent had not been paid the copyhold would have returned to the Corsellis family anyway. Mary's son your cousin John Flack is as much a descendant of Robert and John King as you are. It was your father's father's will!"

"You are right," said I.

"You are my Will," she said with a big smile.

When we had stopped laughing we agreed that our family, our girls and their children are, as Father would have said, "our string of pearls," and are worth far more to us than the lease of the old shipyard.

Sitting round the fire of an evening with Sarah, 'tis only us now, our two girls Mary and Faith are married off. Sarah's face is aglow from the embers, her fingers darting back and forth as she darns my old work jacket. We read to each other from the broadsheet or Father's journal. Reading about his young life we are struck by the losses he suffered. His mother, then his father, and the shipyard! But it was balanced by the love his Aunt Mary and Horace had given him. We discovered just how important Horace sharing his love of books and stories had been. It gave Father a way of standing back and finding meaning in everyday things, a defence and inner strength to endure the pain of loss. But sometimes it was a way of hiding from the world and himself, stowing away the hurt he must have felt from himself and all around him. Sarah says she can see where I get my thoughtfulness from, my way of making the best of things: "Old William gave you the best start in life," she says. I miss his insights and crazy stories.

The broadsheet sparks off our conversation into wider matters, such as another war with France. I tell Sarah about one of the Bermudan sloops old Clear talked about. She was the fastest in the fleet and carried the news of Nelson's death and the victory at Trafalgar, and she was called *HMS Pickle*. I think the name is ironic, the news being bitter-sweet, but Sarah just gives me a look as if to say: another of your random facts! So I have to tell her we have two survivors of Trafalgar living right here in Wivenhoe: Benjamin Snood was aboard *HMS Swiftsure* and James Martin aboard *HMS Neptune*. Then we talk of the success of uncle John's engravings for the King and Sarah tells me of a new book she is reading, written by a woman. "I do not know why you look so surprised at that!" she says. "Women are better at grasping matters of the heart." This author has written many books anonymously. She tells me that the heroine in the book was persuaded by her family not to marry the soldier she had fallen in love with. Both remain unmarried and meet again in later life. She will not tell me any more.

We wonder why my father never remarried. Sarah thinks he was so in love with Mary and lost her so soon that no other woman could replace her. I feel that was probably true and tell her I am sad that I never knew my mother nor saw her with my father. "He loved his work and his books," I say. I still miss him. I confess that after he died, sometimes when I was working I'd turn to tell him: "Oh, that's what you meant..." I'm comforted when Sarah says she sometimes talks to her dead mother too. I tell her he always loved to sit by the river and watch the clouds. I remember how he'd say to me: "Look up Boy! See how small you are under the vast heavens." And I can now understand the relief of feeling small and unimportant.

We have a new cast-iron stove in our fireplace and if we burn coal it will stay alight all night. In the *Ipswich Journal* we read mackerel is plentiful and selling at "1d. to 1½d... very large and fresh." Sarah likes to fry mackerel rolled in crushed oats, such a flavoursome fish. And she bakes rolled-oat honey biscuits. She's never happier than when she's trying out new recipes, whereas I love to do the same thing over and over, refining and perfecting the process. Our two girls reflect both these traits. And Father always said he could see my mother in our youngest.

This year the sea lavender seems brighter on the marsh to our door. As the evening draws on our conversation quietens; we sit in the glow of companionship and the reassuring tock of the clock. The fire crackles and the wind is at the window frames, where the shutters hold back the blackness of the night.

COPYHOLD

Afterword

Part 1

I hope those who are more rigorous about the facts can excuse my liberties of muddling fact with fiction to tell a story of Wivenhoe in the eighteenth century. There was a William King (shipwright) who had a wife called Hannah, but my William King, his son and their wives and family are fictional.

It may have been Francis King's daughter Mary King (spinster) who inherited from the Feedam family and not Mary King (widow) but there is definitely a connection between the widow's daughter who was also a Mary King (and became Mary Flacke) and the shipyard. But there is no mention of the shipyard in William King (shipwright)'s 1723 will.

Mary King (spinster) was Mary Martin's cousin, who wrote in her will of 1703: "I give to my cousin Mary King (4 years old) one Guinea to be paid within three months after my death." Mary Martin's son, Captain Matthew Martin, leaves money "to my cousin Mary King (spinster) who has been living with me," in his will of 1748, as does Matthew's son Samuel in 1765. In her own will in 1773, Mary King (spinster) 66 years old states: "I Mary King late of Wivenhoe but now of Colchester ... desire to be buried in Wivenhoe." She bequeathed everything to Captain Thomas Best and his children with his first wife, her sister, Hannah King. There was a definite need to distinguish between the two Mary Kings at the time with the use of the labels widow or spinster. To reinforce the King/Martin connection Francis King's will mentions his cousin was George Martin (sailmaker), Capt. Matthews Martin's brother. Matthew's daughter Mary married Isaac Lemyng Rebow in 1729. A connection between two of the wealthy Wivenhoe families.

Grandmother Mary King (widow) may have been Mary Robinson who married Thomas King in Ardleigh in 1699. They had one daughter in Ardleigh called Mary in 1700 and perhaps moved to Wivenhoe after her husband died. A Mary King is shown to be in occupation of one of Jonathan Feedam's tenements in his will of 1718. Or it could have been Mary Carter who married Robert King, blockmaker in 1675. They had five children, the youngest a son called William in 1694, but I cannot find a daughter called Mary. If this is our grandmother Mary King (widow) she was born in 1656 and died in 1746, which would have made her 90 years old.

The earliest map of 1738 shows the site of the shipyard alongside Anchor Hill and the graving bridge, in the position of the Anchor Inn. In the Manorial court roll the first reference to a shipyard is in 1575, being worked by John Quixle, spelt many different ways in the parish records. A Court Roll document dated 1507 states: "Richard Quykesly took, for 6d a year rent, half a rood of vacant land next to a dokke of the lord, on the south, abutting west on the land of John Cuttelee called le Werkyng yerd and East on a way leading to le Wherfe".

I interpret this as there being a shipyard east of Quay Street (the road to the wharf) and having a graving dock alongside Anchor Hill. However, the shape of the dock or slipway is the same shape and configuration as in the later 1799 map, which shows the shipyard site upstream of the Woolpack Inn (renamed the Ship at Launch) on the wharf. There may have been a shipwright's place between the row of older, high-pitched thatched houses (including Trinity House) and Anchor Hill, before the Anchor was built (c1684). The Anchor Inn site is at a strange angle to the river andwould match the footprint of a slipway or dock, with the Hard running into the river where the old sailing club hard is now.

Robert and John Page were building ships between c1650-1671, but it may have been too tight a site for the Pages to build a 400-ton vessel as the 1738 map indicates. We may never know for sure.

Documentary evidence is also vague. There does seem to be a connection between a wharf and land called Burrs on which John King had built the house where he was living when he died in 1670. In 1595 a Jacob Ashley, shipwright, refers to "my house and land called Burrs in Wivenhoe with 20 acres of land."

The woodyard on the quay and the the ship and stables (possibly the Ship at Launch) that John Harvey held the copyhold of, with land called the Woodyard (in West Street) where he built a house (c1720s) and where his son later lived, are mentioned in the court books relating to the woodyard transcriped by Pat Marsden on the Wivenhoe history website. It later mentions the wharf and customs warehouse, "formerly in the occupation of Robert King," but no shipyard! I cannot be sure if Robert King was a son or a nephew of Captain John King.

An entry in other deeds dated 6 April 1763 refers to "A certain place called the Old Storehouse Woodyard and Key (sic) formerly in the occupation of Robert King which said premises were lately in the occupation of Captain John Harvey." The shipyard deeds were written nearly eighty years after William King is said to have left the shipyard to his sister Mary King, to justify the Wyatt family's ownership of the shipyard so as to be able to sell it. So it may not be quite correct that the shipyard copyhold ownership came to "Mary, (nee King) wife of Horace Flacke, only sister and heir of William King to all that is commonly called Shipwright's Yard together with the blacksmith's shop." Mary's son John Flacke inherited it from his mother and subleased it to John Iffe, who was in occupation before he "assigns them then to George Wyatt". The only other family connection to the shipyard is Horace Flack's advertising it to let in 1756 just before his death.

Oh, and Robert Clear did not run away until 1750, and everything else about his life is also made up!

Captain John King was married to Mary Parker, daughter of William Parker and sister of John. Captain John Parker was appointed to the *Nonsuch* in 1661 and commanded the 36-gun *Amity* in 1664 at the battle of Lowestoft, and subsequently the *Yarmouth*, a fourth rate of 52 guns, in 1666. "He did not long enjoy his last command. He fell, however, in the hour of victory, being killed in that ever-memorable fight, on the 25th of July, 1666, between the English fleet, and the Dutch." (Pepys)

Part 2

My fictional William and Sarah King help tell the stories around Wivenhoe waterfront and the quay families of Sanford, Stacey, Sainty and Sutton.

There was a Charles and Sarah Stacey with a daughter Sarah but she was born 40 years later than my fictional Sarah.

Thomas Tunmus had a child he named after himself with Mary Lee in 1761. It seems to have been common practice if a wife was barren for a husband to have his children with a surrogate. Nicholas C. Corsellis VI (b.1763) married Mary Bond in 1796. There were no children from that marriage, but Nicholas had 7 children with Sarah Plampin (spinster) who he described as his friend in his will. Sarah was one of the executors of the will.

The new Church was established in Brightlingsea in 1813 but did not really get going in Wivenhoe until about 1864, with a Welsh pastor, David Goydor. The church seems to have had a particular attraction for the shipbuilders Thomas Harvey and James Husk. Thomas would walk about with a long wooden staff looking like some ancient prophet. His son John married Rev. Goydor's daughter and James completely paid for the New Church and Pastor's House at the top of Alma Street. The meeting house could hold 200 people and was lit by gaslight. Thomas Sanford built the Congregational chapel in 1847 with a capacity for six hundred people.

Benjamin Stacey (shipowner) died in 1828 and was buried in the family vault, which still stands in the north-east corner of the churchyard today. There is also an inscription for Benjamin's son Captain Benjamin Stacey. He died in New South Wales. Daniel Stacey of His Majesty's revenue cutter *Repulse* may also be there with his wives and daughter but there is no inscription for them. Perhaps the illegible stone that stands beside Benjamin's tomb is Daniel's. Or perhaps he was buried with his first wife Elizabeth as his will requested which might have been in a family vault in Mistley. There was a John Stace (Custom House Officer) buried in St. Marys Church, Mistley in 1798. And a Stace (widow) from Wivenhoe buried in 1799.

Benjamin Stacey's will of 1828 mentions three ships: *New Union*, *Preston* and *Fox* (perhaps Daniel Sutton's lugger *Fox*) along with his dwelling house and granary on Wivenhoe Quay and also the newly-built brick "Admiralty Warehouse" in Brightlingsea. After Benjamin's death three of his vessels were auctioned at Billingsgate: the *Welcome Messenger*, burden 43 tons, the *Fox*, burden 40 tons. "The *Fox* is a fast superior sailing vessel having been built by the celebrated Mr. Sainty" (*Ipswich Journal*) and the *Nancy*, 39 ton or thereabouts, sold by John Green Chamberlain (Officer of Customs). John was the husband of Benjamin's daughter Catherine Stacey. After her death John married Daniel Stacey's daughter Elizabeth Sarah and their son is also called John Green Chamberlain. (Both John's wives die after 14-month illnesses. I wonder if there was arsenic wallpaper in the house they shared.) John Green Chamberlain the younger later became a merchant, Deputy Sergeant of the Cinque Ports and agent for Lloyds at Wivenhoe, and lived on the quay in the Stacey's old house until he built a big house called the Nook in 1864.

Philip John Sainty (1753-1844) took over the Wivenhoe yard from William Hawkins (1759-1812). The largest of four ships was *Pearl*, a 558-ton, 18-gun sloop of war for the Royal Navy. March 3, 1828 records the launch of the 600-ton, three-masted sloop of war. She stuck on the ways during launching and was not got off until the next high tides a fortnight later. *Cupid* (1829) was a 202-ton, 78ft long snow; *William & Mary* (1831) a 296-ton, 101ft long barque; *Venus* (1832) a 159-ton, 71ft schooner; *Fancy Lass* (1833) a 105-ton Brigantine. All four for William Hawkins (II) timber merchant at the Hythe.

Thomas Harvey (1803-1885) took over the yard from Philip Sainty after he was made bankrupt in 1834. Harvey's first big patron was a consortium, with the Sanford family naming the schooners *Lady Rebow*, *Gurdon Rebow* and *Slater Rebow*. A fourth schooner was named *General Rebow* for two different owners. Thomas also built two more yachts for Lord Paget's, family. Harvey's Wivenhoe shipyard suffered a severe fire in 1872 and did not fully recover. Harvey built some very fast cutters and schooners and after the schooner America beat his Ipswich-built cutter *Volante* in 1851 everyone wanted schooners, and the America's Cup race was born.

From the days of the fast revenue cutters and Sanity's and the Marquis' *Pearl*, the Colne has had the reputation for building fast and manoeuvrable cutters, smacks, cod smacks, fruit schooners and yachts. This followed through into the building of luxury steam yachts for the wealthy Victorians.

Bibliography & Thanks

Websites:
Wivenhoe History: wivenhoehistory.org.uk
Public Records Office: nationalarchives.gov.uk
Essex Records Office: essexarchivesonline.co.uk
Parish Records Online: essexandsuffolksurnames.co.uk
Ipswich Journal: britishnewspaperarchive.co.uk
Warships in the Age of Sail: www.threedecks.org

Literature:
Hervey Benham: *The Salvagers* (Essex County Newspapers Ltd, 1980)
The Smuggler's Century (Essex Record Office, 1986)
Essex Gold (Essex Record Office, 1993)
David Patient: *One of Howard's* (Jardine Press Ltd 2020)
Nicholas Butler: *The Story of Wivenhoe* (Quentin Books, 1989)
J Dodds & J Collins: *River Colne Shipbuilders* (Jardine Press 2009)
J Dodds & J Moore: *Building the Wooden Fighting Ship*
(Hutchinson & Co, 1984)
Tre Tryckare: *The Lore of Ships* (Holt Rinehart & Winston 1963)

Special thanks to:
Catherine Dodds, editor and designer
David Patient, for advice and encouragement
Des Pawson, for knot and rope expertise
Edith Webb, for costume advice
Frances Belsham & Pat Marsden for their excellent work on the
Wivenhoe History site
Belinda Bamber, for copy editing

Timeline (Wills and key dates) references

1606 Thomas King marries Mary Gray	
1636 William Parker dies	Will D/ACW 12/136
1665 Mary King marries Capt. James Lock	
1666 John Parker killed in action	
1675 James Lock dies	D/ACW 18/346
1670 Capt. John King dies	Will D/Y37/2/129
1673 Willliam King marries Ann Feedham	
1675 Susan King marries Capt. William Parker (jnr)	
1675 Robert King marries Mary Carter	D/AEL1767/26
1679 John Page dies	Will D/ACW 19/189
1686 Francis King dies	PROB 11/396
1656 Grandmother Mary King born	
1694 Mary King (widow) dies (Mary Martin's mother)	Will D/ABW 75/65
1690 Aunt Mary born	
1699 Robert King dies	
1714 King George I	
1714 William King Born	
1716 James Feedham dies	D/ACW 24/225
1718 Jonathan Feedham dies	T/G 383/3
1720 Southsea bubble crash	
1722 Horace Flack marries Mary King	D/ABL 1722/32
1726 William King senior dies	
1727 King George II	
1727 Apprenticed to John Iffe	
1730 Francis King (Jnr.) dies	PRO 11/247/43
1735 Finishes apprenticeship	
1735 William King marries Mary Hall	
1739 Aunt Mary Flack dies aged 49	
1741 Horace Flack marries Leah Feedham	D/ACL 1741/15
1740 Robert Hopkins dies & Michael Hopkins born.	Will PROB 11/705
1743 Young Will King born	

1743 William's wife Mary dies aged 30
1745 John Davis dies (last apprentice James Harvey taken on 1744)
1746 Grand Mother Mary King (widow) dies Will D/ACW 29/7/3
1746 Austin Stapley marries Mary Roberson (or was it Austin jnr)
1746 William Halls dies Will T/G 383/5
1755 Build sloop with son aged twelve
1756 Horace Flack dies aged 58 Will D/DE1 T190
1757 Tomas Tunmer marries Anne Flack D/ALL1757/181
1757 Young Will apprenticed to John Iffe
1760 Anne Tunmer dies
1760 King George III
1762 Thomas Tunmer and Hannah Cole marry D/ALL1762/137
1763 New dredging laws
1764 Michael Hopkins dies Will PROB 11/897
1764 Will marries Sarah Stacey
1769 John Iffe dies aged 67
1773 Mary King (spinster) dies Will D/ACW33/3/14
1776 George Wyatt dies aged 64 Will D/ACW33/6/28
1776 William King dies aged 62
1779 John Sanford dies aged 64 Will
1794 John Green Chamberlain born
1805 Elizabeth Sarah Stacey born
1805 Battle of Trafalgar
1808 Capt. Daniel Stacey dies age 44 Will D/DEtF10
1808 William Stuttle dies aged 68 D/ACW38/4/24
1815 Joseph Cole dies aged 66 D/ACR 19/672
1816 John Constable's painting of Wivenhoe Park
1817 William Hawkins dies aged 53 D/ACW39/2/33
1820 King George III dies
1820 Launch of the *Pearl*
1821 Will King dies aged 78
1828 Benjamin Stacey dies aged 65 Will D/ACW41/5/16
1854 Philip John Sainty dies aged 90

Fictional characters and dates in italic